The Miner's Freedom

WORK

Its Rewards
and
Discontents

Advisory Editor

Leon Stein

THE
MINER'S FREEDOM

CARTER GOODRICH

ARNO PRESS

A New York Times Company

New York / 1977

Editorial Supervision: ANDREA HICKS

———————◆———

Reprint Edition 1977 by Arno Press Inc.

Reprinted from a copy in
   The University of Illinois Library

WORK: Its Rewards and Discontents
ISBN for complete set: 0-405-10150-3
See last pages of this volume for titles.

Manufactured in the United States of America

———————◆———

Library of Congress Cataloging in Publication Data

Goodrich, Carter, 1897-1971.
   The miner's freedom.

   (Work, its rewards and discontents)
   Reprint of the 1925 ed. published by M. Jones,
Boston, in series: The Amherst books, second series.
   Includes index.
   1. Coal miners--United States. I. Title.
II. Series. III. Series: The Amherst books, second
series.
HD8039.M62U626   1977   331.7'62'2330973   77-70498
ISBN 0-405-10169-4

The Amherst Books
Second Series

THE MINER'S FREEDOM

Volume Number 1

ON THE AMHERST MEMORIAL FELLOWSHIPS FOUNDATION

THE AMHERST MEMORIAL FELLOWSHIPS FOR THE STUDY OF SOCIAL, ECONOMIC AND POLITICAL INSTITUTIONS were founded in 1920 by an anonymous graduate of Amherst College to perpetuate the memory of those Amherst men who gave their lives for an ideal. The donor stated their purpose as follows:

"Realizing the need for better understanding and more complete adjustment between man and existing social, economic, and political institutions, it is my desire to establish a Fellowship for the study of the principles underlying these human relationships."

THE MINER AND HIS PLACE

Reproduced by permission of The Survey-Graphic.

Photo by Lewis W. Hine.

THE
MINER'S FREEDOM

A STUDY OF THE WORKING LIFE IN A
CHANGING INDUSTRY

BY

CARTER GOODRICH

ASSISTANT PROFESSOR OF ECONOMICS IN THE
UNIVERSITY OF MICHIGAN; AMHERST MEMO-
RIAL FELLOW IN ECONOMICS, 1921-22, 1923-24;
AUTHOR OF "THE FRONTIER OF CONTROL"

BOSTON
MARSHALL JONES COMPANY
1925

Printed in the United States of America

THE COLONIAL PRESS
C. H. SIMONDS CO., BOSTON, U. S. A.

ACKNOWLEDGMENTS

F OR the freedom and leisure and field travel that have made this book possible, I am indebted to the very generous provisions of the Amherst Memorial Fellowships. The Bureau of Industrial Research also contributed a part of the expenses of the first year of the investigation, and I owe a number of valuable suggestions in the planning of the study to those who were then associated in its work—and particularly to the late Arthur Gleason and to Robert W. Bruère and Heber Blankenhorn—as well as to the Fellowships Committee.

The most useful single set of documents to which I have had access is the collection of union records and of joint conference and arbitration proceedings preserved by Secretary Richard Gilbert in the district headquarters of the United Mine Workers at Clearfield, Pennsylvania. But on a problem which called for the knowledge of customs and practices and processes as well as of contract provisions, I have gained less from any documents than from contacts with the day-to-day life of the industry,—from the talk of miners and engineers and operating men, from the public discussions and private conversations at union conventions and at meetings of operators, and from the days that a number of companies have permitted me to spend underground. Both outside students of the industry and men actively engaged in its work and in its controversies have been extraordinarily willing to talk over its problems and to open up channels of information, and my obligations to them are far too many to mention. I remember with particular pleasure, however, the friendliness of the union organizers who took me along on some of their dashes into non-union Somerset early in the 1922 strike and of the foremen and "face bosses" who have taken me on their rounds in a dozen mines. Among government experts, also, I owe especial gratitude to Mr. F. G. Tryon of the United States Geological Survey; among mining engineers to Mr. Hugh Archbald, author of "The Four-Hour Day in Coal;" among operating men to Colonel William M. Wiley of the Boone County Coal Corporation and to Colonel Edward O'Toole of the United States Coal and Coke Company; and among people active in the union movement to President John Brophy of District Two and many of his associates and to Edward A. Wieck and Agnes Burns Wieck of Belleville, Illinois. It is needless to say that no single statement about the bituminous coal industry could be acceptable to all of these people, and I am sure that no one of them would care to take responsibility for my viewpoint in its entirety.

Many friends have been kind enough to give advice during the work of writing, but the manuscript owes most of all to the helpful suggestions and incisive criticism of Florence N. Goodrich, Walton H. Hamilton and Helen R. Wright.

C. G.

Ann Arbor, Michigan,
   December 1, 1924.

CONTENTS

ILLUSTRATIONS

THE MINER'S FREEDOM

PART I

THE MINER'S FREEDOM

CHAPTER I

INTRODUCTION

IT IS often said that modern society has chosen efficiency in production rather than richness in the working life; that it has chosen the possibility of fuller and more varied living outside working hours rather than the possibility of a creative life on the job itself. The statement is as often made by those who regret the "decision" as by those who glory in it, but in either case it is nonsense and perhaps dangerous nonsense. It may be a fair enough description of what we seem to be getting (at least aside from the question whether the people who do the more varied consuming and those who do the more monotonous producing are always the same); but to say that we chose it and that we chose it once and for all is not only a false rationalization of the past but a real hindrance to intelligent thought about the future. Society makes no choices as such, and the countless individual decisions out of which have come mass production as efficient as that at Ford's and jobs as dull as those at Ford's have most of them been made without the slightest reference to the quality of the working life that would result.

Changes in the way that work is done do not come by plebiscites or by single dramatic choices. The immediate decisions rest typically with individual business men acting upon the reports of their hired technicians. And no matter how revolutionary their social consequences may be, the decisions themselves are certainly not made on revolutionary grounds. They are made on the basis of figures of output and cost and profit for the immediate business in the immediate future. Vaguer factors of prejudice and sentiment and social philosophy no doubt enter in more often than the strict logic of business demands, but even among these there has been little place for a desire to make the worker's job either more or less interesting for its own sake. Surely no one has slain craftsmanship out of malice, and the case is almost as clear with those changes in technology that have done away with some of the worst forms of brute tugging and hauling. Here very likely the technicians have often worked with additional enthusiasm, but even those engineers who are happiest in the feeling that their innovations have improved the quality of men's work tell of it as a "glorious" accident. "Whenever we start up some bold and economical change in our work," writes one of them, "we almost invariably find collateral, incidental and unsuspected benefits as great as those we had first in mind."* These changes in technique have been altering the whole nature of human work, but their effects on the working life, good and bad, have been mere by-products quite "unsuspected" by men who have planned them carefully for other ends.

This is only part of the story, to be sure. While it is the technicians who invent the new jobs and the employers that offer them, it is the workers who decide whether or not they will accept them. But it is obvious that these decisions are made on the basis of the opportunities that are open and of which the workers happen to know, and on an income level that permits few luxuries. Perhaps the wonder is that under the circumstances the preference for one kind of work over another enters as often as it does into decisions dominated by the imperious need of wages; but surely no realistic student of the actual alternatives is likely to think that these preferences have very actively determined the course of industrial development. The choice between dull job and no job, for example, is an easy one to make, but it exerts no selective influence.

For the organized workers the situation has no doubt been somewhat different. The unions have often forced some consideration of their in-

*Arthur J. Mason, "Comments of an Ore Engineeer," *American Economic Review Supplement,* March, 1921, p. 109.

terests into the decisions on technical changes. They have often delayed them, they have frequently bargained over the conditions of their introduction, and they have sometimes stood out against them altogether. And part of the driving force of these policies, whatever the public statements of the leaders may have been, has often come from the stubborn liking of the members for their old kind of work or stubborn prejudice against the new. But clearly the main object has always been the defense of wage rates and of the employment that was threatened by the change, and what concessions they have secured have been in terms of wages and of special claims to the new jobs and not in terms of technical adjustments to make the new jobs more attractive in themselves. The union that opposes a technical change too vigorously, moreover, is far less likely to prevent the change than to eliminate itself. Whatever their desires may have been, neither organized nor unorganized workers have been in position to press any vigorous claims to a right to workmanship; the right to work of any kind has been too precarious.

In all of the immediate decisions, then, there is no sign that the weighing of the desirability of one sort of working life as against another has played any decisive part. And this is true also of the general public attitude against the background of which these choices are made and by which they are largely affected. Technology grows where it is encouraged, and even in the detailed decisions the participants keep something of a weather eye for government and "public." Union leaders realize that they cannot openly oppose technical changes without the certainty of public condemnation; and when the coal operators, for example, declare that it is the duty of the worker to "lend himself without reserve" to whatever technical innovations they may choose to impose, they can be confident that most of their readers will take it as a truism. For the belief in technical progress without thought of its possible costs is one of the most widespread and most important of the preconceptions that make up the current point of view. This faith has many grounds; the very brilliance of the technicians' achievements makes it natural enough; but it is clear that neither in lay thought nor even in economic theory is it based on any reasoned comparisons between one sort of work and another. For economics has always been in its major emphasis a consumers' science, in the sense that it has taken consumption as the end of all things and labor as merely a disagreeable means; and one dominant branch of theory tends to obscure the differences between one kind of work and another by concentrating upon the point at which all labor can be taken as "one kind of pain."* Even labor legislation has touched the actual working life only at points of safety and sanitation and the like; and neither in general thought nor in the specific decisions has there been any careful weighing beforehand of the gains and losses in creative work in the same scale with the gains in the output of the consumers' goods. There has been no choice, that is, in terms of the quality of the working life.

No choice; yet a time when the working life has been changing more rapidly than ever before in the history of the world, and so radically that observers of some of the new processes declare that what is going on "does not seem to be labor" at all but something "fascinating and horrible" and so different from anything in the past that it should be given an entirely new name!† The very magnitude of these changes is at last forcing them on our attention, and a number of modern voices are renewing the unheeded attacks of Ruskin and Morris against the kind of

*J. B. Clark, *The Distribution of Wealth*, p. 380. Perhaps the most realistic form of the argument is that to be found in Alfred Marshall's *Principles of Economics*, sixth edition, Book IV, Ch. I, Sec. 2. Note as an exception to the general trend Adam Smith's discussion of the effect of "their ordinary employments" on the "understandings" of common men in Book V of *The Wealth of Nations*.

†Robert Littell, "Henry Ford: Men and Machines," *New Republic*, Vol. 36, p. 222.

work that turns men into mere machines. I do not argue that this outcry is always justified; not all of the unplanned changes in the working life have been changes for the worse, and even the losses may often have been offset by gains in life outside the job. And of course I do not argue that there need be a blanket choice between technical advance on the one hand and the possibility of workmanship on the other; both are desirable, and the problem is one of adjustments rather than wholesale renunciations. But I do argue that the issues raised by this new *producer's protest* are vitally important—for men are alive (or should be) even during working hours—and that the issue of the quality of the working life is too important to be left to chance either in the devising or in the adoption of the new technique. My argument is not that we have always chosen badly but that we have not chosen at all, and that we should find ways of forcing this issue into consideration, not in sentimental regrets after the event, but in the actual industrial decisions themselves.

But choices imply a knowledge of alternatives, and the first essential in any such program of choice (and the only one with which this study is directly concerned) must be the picturing beforehand of the different types of working life which the change involves. In the past this has been largely impossible. The immediate pair of jobs could be compared after a fashion, of course; but no one, and the workers least of all, could foresee where a given set of changes was likely to lead. In this as in other matters, it is true of the first coming of the machine that "by the time men began to see its nature, to discover its effects, and to weigh its consequences, (its) influence had permeated the whole social system and its effects had gone to the ends of the earth."* But it is the hope of social scientists that this need not be so completely true of the continuing industrial revolution of our own times. We can perhaps do little in the way of confident prediction in regard to the intensification of the machine process that is going on so rapidly in American mass production, even in a matter as immediate as its effect on the working life, though we should certainly be aware of the problem and have some tools for handling it. But in the matter of the spread of the machine process to industrially backward nations and regions and to "backward" industries in industrial nations, the problem of prediction ought surely be easier. The march of the machine around the world is perhaps the most dramatic movement of our time; its steady spread from in-

---

*From an unpublished manuscript by Walton H. Hamilton on the theme that "the lords of rising industrialism were not revolutionists at heart."

dustry to industry and from process to process within our own country is less picturesque but scarcely less significant. These changes, for all the rich complexity of each individual case, are after all primarily the carrying to new places of old ways of working whose effects we have had time to study; surely in these cases it should be possible for social science to turn its studies of the past into instruments of prediction and guidance. Surely it ought not to be too difficult to give to those who are thinking seriously about the coming of industrialism to China, for example, or to North Carolina, or about the spread of machine methods to the coal industry, some genuine indications of the social as well as the immediate technical consequences of the change. And among these it is clearly possible to give some picture of the alternative types of working life.

It is with the last problem as it appears in the bituminous coal industry that the present study is concerned. It may be paradoxical to speak of a current coming of industrialism to coal, since without coal there would have been no industrialism. Yet it is true that, so far as the manner of work underground is concerned, "mining is still in a way a 'cottage' industry." The *indiscipline* of the mines is far out of line with the *new discipline* of the modern factories;

the miner's freedom from supervision is at the opposite extreme from the carefully ordered and regimented work of the modern machine-feeder. The contrast is sharp and striking, and moreover it is not merely a static one. Machine methods of production and factory methods of supervision have already begun to invade the mines, and the industry is apparently in the first stages of a great industrial revolution. This contrast and this impending change are the subject-matter of my study. Those who are deciding to change the industry for other purposes are incidentally voting a different sort of life to the miner. My purpose is to picture the contrasting types of working life as vividly and as accurately as possible in the attempt to bring these values also within the range of judgment and current decision.

CHAPTER II

THE JOBS OF THE MINE WORKERS

"THERE'S no such thing as discipline in a coal mine." "The miner is his own boss."—Neither of the two sayings is literally true, but both are current and together they point to a real and important contrast between the working life inside the bituminous coal mines and that in modern factory industries.

The fact of this difference is most often and most vigorously insisted upon by the engineers and other technical enthusiasts who are anxious to make a mine "more like a shop." It is they who speak most often of "the peculiar psychology underground" or of the "old-world mental attitude prevailing in the rooms of our coal mines"* and of the "hit-or-miss supervision"† from which the miner himself is said to suffer most of all. A quite different clue to the quality of this life comes from a colored loader in a West Virginia mine. On the occasion of the foreman's daily visit to his place, he had stopped work long enough to explain why he, like so many other

*Mason, *op. cit.*, pp. 107, 109.
†United States Coal Commission (1923), *Underground Management in Bituminous Mines*. By Sanford E. Thompson and Associates. P. 2. (This document will be cited as the "Thompson Report.")

miners, had drifted back from factory work to coal mining. In the mines, he said, "they don't *bother* you none." And still another indication comes from the records of a union convention. A proposed clause in the joint agreement was under discussion which "recommended and requested" that the men stay in the mine the full eight hours "for the mutual benefit of both." "That's taking away a little freedom from a man, hain't it?" asked one of the delegates. "I think a man ought to know when he is tired."*
And the clause was defeated. Such a rule, to be sure, has long since been written into the union contracts, but the delegate's question is still close to the spirit and practice of the mines.

The more public pronouncements of the union naturally say little about the unique character of the work, although in 1920 a high official of the international organization made an eloquent appeal to "the traditions that have grown up with the industry" in an argument against "the installation of time clocks and those other new innovations." The statements of the operators, however, refer more often to the peculiarity. Sometimes it figures in the argument as a reason why miners should be contented, on the ground that "there is no other industry in which work-

*United Mine Workers of America, District 2 (Central Pennsylvania), *Convention Proceedings*, 1904, Vol. IV, p. 32.

men enjoy, to such an unusual degree, entire freedom of action and untrammeled privilege of working when and as they please." But more often the operators use it in bitter complaints that their rules are "treated with impunity"* by miners over whom they have no control. In private conversation, however, they are quite likely to speak of the fact that the miner is a "sort of free lance" with resignation or even tolerance. "The miners have been independent since the year one, and I suppose they always will be."

Outsiders, too, have noticed the peculiar psychology of the industry, and a succession of them have echoed in one form or another the remark of an eighteenth century observer that miners are "greatly impatient of controul."† A striking corroboration of this comes recently in the form of a warning addressed to employers in other industries. Don't hire former miners if you can avoid it, advises a writer in *Industrial Management*. Their work has unfitted them for factory discipline. In the mines, "the possibility of constant supervision or of surprise tests does not exist. The coal miner is accordingly trained to do as he pleases. . . . Transplant

*Bituminous Coal Commission (1920), *Transcript of Testimony*, Vol. VI, pp. 296, 252, 482.
†Arthur Young, *Northern Tour*, ii, 261 (1768). (Quoted in J. L. and Barbara Hammond, *The Skilled Labourer*, p. 19.)

such a man into a factory where production is speeded and no imagination is required to picture what will happen." His "personality" is "so magnified" by the laxness of mine discipline "that he chafes under the necessary restrictions of other employment . . . . The ex-miner resents all suggestions as to his working methods, resents all effort to compel continuous application, and assumes in general a hostile attitude to all supervision."*

Some of these statements are exaggerated; in their judgments of good and bad they run the gamut from the hint that modern industry is unfit for miners to the round declaration that miners are unfit for modern industry; but they all point to the same sharp contrast. It is clear that the working arrangements out of which these attitudes arise and to which these comments apply are far out of line with those that prevail in modern machine industry. Call it freedom or call it indiscipline: this peculiar characteristic of the miner's work is worthy of study by anyone who is looking critically either at the coal industry or at modern industrialism, and who believes that the quality of the working life within it is one of the important criteria by which an industry is to be judged.

*H. A. Haring, "Three Classes of Labor to Avoid," *Industrial Management*, Dec., 1921, p. 370 *et seq.*

The Men at the Face

If it is true that "the miner is a little peculiar," as an apologetic operator once told a government commission, it is largely mining that makes him so. For it is not the miners who decide how the mines shall be laid out or how the industry shall be organized, and the characteristic indiscipline is deeply rooted in the prevailing technique. "Mining is still in a way a 'cottage' industry," as the Editor of *Coal Age* has pointed out, "only the cottage is a room in the mines;"* and the miner's freedom is largely a by-product of the very geography of the working places inside a mine.

It is with these *rooms*, then, and with the other *places* at the *working face* that a description of the jobs of the mine workers may well begin.   For this is where the basic work is done; it is here that the coal is extracted from the solid seam and shoveled into mine cars. The rest of the labor of the mines is mainly transportation of the coal or some form of "service to the men at the face." The face corresponds quite accurately to the front in the organization of an army at war; the other workers are thought of as forming "a service of supply." The men at

*R. Dawson Hall, "Have Mining Engineers Accepted All That Developments in Machinery for Handling Coal Imply?" *Coal Age*, July 7, 1921.

the face, pick miners and loaders, make up over half of the total number of employees; and it is their work and their relation to the isolated places inside the mine that give much of the characteristic color to the discipline situation in the industry.

Isolated places, for at one significant point the army analogy fails to hold. The working face is not an unbroken line; it is in fact the outstanding characteristic of the prevailing room-and-pillar system of mining that the face is broken up into scores of scattered faces, each belonging to a separate place. And each place belongs, "almost as a personal possession," to a pair of miners, working as "buddies," or perhaps to a single miner. Once a man has started work in a place, he is rarely transferred from it until it is finished, unless he himself asks for a better one, and it is often held for him a considerable time when he is away. The miner is first of all a workman in a unique relation to the working place which he is said, with little exaggeration, to "own and inhabit."

But what *is* a place, and what does it look like? It is typically a narrow tunnel, somewhere inside a mine, that is being pushed forward into the coal itself. The inside end of it, the face from which the coal is mined, is advanced a few feet a day. The height is usually that of the seam of coal itself; more often than not a man cannot stand upright. If the place is a room, the width is perhaps twenty-four feet. On either side are solid walls of coal, called pillars, which are left to support the roof. Not far from the face, these walls are broken by narrower tunnels, cross-cuts or break-throughs, that lead to the neighboring rooms and provide for ventilation. Aside from these openings there is nothing to suggest a connection with the rest of the mine except the narrow-gauge track that runs the length of the room and out into the heading, or haulageway, beyond. Up this track come the empty cars to be filled; down this track go the loaded cars. This is the chief contact with the service of supply; the only contact with the company is the infrequent visit of the mine foreman or his assistant.

If the place is not technically a room, the description differs in details but hardly in the main conditions that affect the miner's freedom. If it is a heading or an air-course, or if the miner has turned aside to drive a break-through, the place is simply a somewhat narrower tunnel. There is a real difference in the look of the place only if the miner is "pulling pillars," that is to say, taking out the coal that was left behind to hold up the roof when the rooms were first driven up. Here he is no longer working between solid walls of coal, and there is more danger to guard against and a need either for more experienced miners or for somewhat more supervision. But in any case the isolation and the independence remain very much the same, and the miner and his buddy work month after month, pressing forward slowly into the seam of coal by the light of the lamps on their caps, "left alone," except for the track and the boss's brief visits, "to handle the inside of the planet as they find it."*

In this isolation—whether in room or heading or pillar—it is clear that "operations at the face" must very largely be "left to the judgment of the miner himself."† But how much judgment, and how much skill and dexterity, do these operations require? "No great brains," is the usual answer from operator or engineer, and no great skill, except perhaps in the part of the industry where hand mining still prevails. "The old miner was a craftsman and a wonderful craftsman," an operator may say, but the loader of today "needs nothing but a strong back." These answers, however, need to be checked carefully against a description of the actual operations. It is true that mining is on the whole relatively

*Heber Blankenhorn, "The Conquest of Isolation," *Survey*, Vol. XVII, p. 1007.
†Thomas A. Stroup, "Causes and Growth of Unionism Among the Coal Miners," *Mining and Metallurgy*, Sept., 1923.

rough and simple work.   It is true also that it is
now for the most part simpler and slightly less
varied than in the earlier stages of the industry's
development, although if the old-time miner was
a wonderful craftsman it was less because of his
dexterity with the pick than because of the
knowledge of roof and gas that he needed in the
days when the boss might not visit his place
from the time that it was started until it was
finished.   But under each of the three methods
of mining that are common to-day, the miner's
work is still much more than "mere shoveling;"
and the examination of the other elements in the
job will make clear the fact of its variety and
will throw some light also on the questions of
skill and judgment.

For almost a fifth of the country's tonnage,
the miner's cycle of work still begins with the
job of undercutting the coal by hand.   To do
this he hacks away at the bottom of the seam
with a short sharp pick, cutting a "V" four feet
deep under the solid coal.   When the cut is part
way in, he places a short piece of timber, a sprag,
under the edge of the coal so that the overhanging mass will not fall on him as he lies under-
neath and finishes the job.   Hard, awkward
work, certainly, though it is said that in many a
team of father and son working together in the
mines, the old man who cannot stand much

shoveling does this part of the work and leaves most of the loading to the sturdier and less skilled youngster.

In a still smaller part of the industry, the places are not cut at all, and the coal is "shot off the solid" with deeper holes and heavier charges of powder. In any form of mining, the blasting of the coal calls for a degree of judgment but in the solid work, without the help of a cut to make the coal fall more easily, the decisions as to the placing of the holes, their depth and direction, and the amount of powder, are matters of special delicacy. Even the planning of a single shot so as to bring down a satisfactory quantity of coal requires considerable knowledge of the "cleat" of the coal and the way it will break, and the skilled miner always arranges one day's shots so as to leave the face of the coal in such a shape that it will be easy to make the next day's shooting effective also. It is in this work, apparently, that the greatest tradition of individual skill and the greatest pride of craftsmanship survive, and it is a well established custom among the solid miners to visit each other's places to see the results of yesterday's shots and to discuss and criticize the placing of to-day's.

Both pick mining and solid shooting, however, are rapidly giving place to machine mining, in which the cutting machine does the work of the

DRILLING FOR A SHOT

Reproduced by permission of The Survey-Graphic.

Photo by Lewis W. Hine.

pick; and now almost two thirds of the country's output is mined in this way. The machine, however, takes over only the much smaller part of the old-time miner's work, and the change has altered the life in the room surprisingly little. The work of the machine runners will be discussed later; the present interest is in the job of the machine loader, or "loader after the machine," who is left to do the rest of the work in the place. The description may well pick up the cycle of his operations just after the cutting is finished. The machine leaves behind it a pile of fine coal that it has ground out of the cut. The loader must shovel this "bug dust" into a mine car before he can lay the pair of rails extending the track to a new face and before it is safe for him to fire his shots. He then decides where to drill his holes for the blasting and at what angle to slant them in order to break the coal down most effectively. The drilling itself is done with a long auger, which the miner braces against his body—perhaps in low coal bracing his back against the roof—and turns by hand.* Then he fills a cartridge with powder, places it in one of the holes, tamps the hole with clay, and finally lights the fuse and takes refuge in the heading or the nearest break-through. When he returns

*In somewhat harder coal, the auger is replaced by the more complicated grip-bar drill.

to the place after the explosion and after some of the smoke has cleared away, he finds part of the coal broken down on the floor and more of it shattered but still in place. The loose coal is ready to be shoveled into the car; the standing coal must first be "pulled" with a pick. From this point on, the job is mainly shoveling,— "mere shoveling," if you like, although in the process the loader must also clean the coal by throwing aside at least the larger pieces of slate and other impurities, and although shoveling itself may involve a considerable knack under underground conditions,—particularly in the many cases where it is necessary, as it was for the miner whose skill was so much admired by the Coal Commission's efficiency experts, to "throw shovel after shovel of coal against the roof from which it caroms into the car."* This cycle of shooting, picking, cleaning and loading is repeated until the place is "cleaned up" several days later and ready for the next cut. But at any point in the cycle the miner may have to stop to take down a piece of slate that is hanging dangerously over his head or to support the roof by setting a prop,—a timber that he sets upright and makes firm by driving a wedge-shaped cap-piece between its top and the roof. For scale contracts and mining laws alike restate the cus-

*Thompson Report, p. 33.

tomary rule of the mines that the miner is largely responsible for his own safety, and it is literally true that "his life depends on where he plants his props."*

Here in the bare essentials of the ordinary loader's job are clearly both variety and the need for continual small decisions, but the actual list of operations is likely to be even longer and more varied. The average proportion of the miner's day that is devoted to each process is unknown, but the division of his time may at least be illustrated from data based on the extensive time-studies made by the United States Coal and Coke Company at its West Virginia mines in 1911 and 1912. The studies of the two years were not made with precisely the same classifications, but they are easily comparable, and between them they cover observations of the work of nearly two hundred and fifty loaders. For the present purpose the times spent on each part of the work are expressed as percentages, not of the nominal working day or of the total time spent in the working place, but of the shorter time spent in actual work.†

The table follows:

*Blankenhorn, *op. cit.*, p. 1007.
†The original data were made available by the courtesy of Colonel Edward O'Toole, General Superintendent, and of Mr. Howard N. Eavenson, formerly Chief Engineer.

THE WORK OF THE LOADER

Operation	Percentage of Total Working Time		
	1911 Study	1912 Study	
Loading Coal	65.2	55.8	51.2
Picking Down Coal			4.6
Re-shoveling Coal		3.4	
Cleaning Coal	4.0	9.6	
Drilling Holes			5.0
Charging Holes		9.9	4.5
Wiring for Shots			.4
Drilling, Charging & Shooting	11.3		
Tracklaying	6.9	5.0	
Timbering		7.9	7.5
Unloading Ties & Props			.4
Unloading and Setting Timbers	6.6		
Taking Down Slate & Rash	1.5	1.2	
Taking Up Bottom	.2	1.8	2.7
Cleaning Out Cut & Place	1.6		
Putting on Wrecked Cars	.8		
Dropping & Placing Cars	1.0		
Pushing Cars to Face		1.0	
Miscellaneous (including Examining Roof & Charging Lamps)	.8	3.4	
	99.9	99.9	

The striking things about the table are the facts that loading makes up little more than half of the loader's job and that the list of other operations is so long. There is no claim that these figures are strictly representative of the industry as a whole. The time spent in setting props, for example, varies widely both with physical conditions and with the care exercised by the workman or enforced by the company, and so with each of the other items on the list.

And the operations themselves vary somewhat from company to company and from region to region. The item of wiring for shots, for example, is peculiar to this particular group of mines; here the firing is done with an electric battery by the assistant foreman. This is quite unusual, but in many regions the shots are all fired after the miners have left the mine by a special group of shot-firers. In some districts the men are required to push the loaded cars away from the face as well as the empty cars to the face, and the records of union conventions are full of vigorous denunciations of the practice and of declarations that the miners will not be "mules" any more;* many mines, on the other hand, have done away with car pushing altogether. But the item, "putting wrecked cars on the track," is characteristic, not of a single company but of the industry as a whole; and another heading, that appears in the day's work of one of the three miners in the Coal Commission's time studies, "advice to other miner,"† is perhaps equally characteristic. The ordinary list of operations would hardly be shorter, and it is probable that the average proportion of time spent in shoveling alone would hardly be greater, than in the illustration given. The table

*See for example the long discussion in the District 2 *Convention Proceedings* for 1917.
†*Thompson Report,* chart 20.

gives a fair impression of the variety of the loader's work and a suggestion of the great number of decisions for which the day's work calls.

But how is it that the company can afford to leave so many of these decisions to the miner? There are several answers to the question; one modern one, as Part II will show, is that the company can *not* afford it; but the chief and obvious answer is that the greater part of the immediate cost of any mistake falls not on the company but on the miner himself. It is the system of payment, as the Thompson Report points out, that relieves the foreman of "the necessity of driving the men."* The miner is a piece worker, paid usually by the ton, and a piece worker in an industry in which the overhead is relatively small, so that it is his own living (as well as his life), and not the company's, that is most immediately affected by what he does or fails to do in the place. If he is clumsy in handling his shovel or his pick, or if he places his shots without due regard to the way the coal will naturally break, it means simply more work for the same pay. And if he lays off for a few days, the immediate loss, whatever the secondary effects on the organization of the mine may be, is borne by the miner himself. As a result it seems to be very frequently the attitude of the industry that

*Ibid, p. 4.

"production is in the main the look-out of the miner." The weigh-sheets and payrolls that record his output (or perhaps his output and half his son's under the same check number) are almost never scrutinized to serve as a basis for discharge; and although the operators often complain bitterly of the excessive absenteeism of the men at the face, actual records of the attendance are seldom kept. A slow or an irregular worker may perhaps be given a poor place to work in or a place whose advancement is not essential to the development of the mine; but the man at the face is rarely discharged for either of these causes. There is in fact a strong feeling in the industry, among bosses as well as among the workers, that the miner is a sort of independent petty contractor and that how much he works and when are more his own affair than the company's.

As the basis of this attitude, then, the fact of piece work is an important part of the explanation of the miner's freedom; and the same fact, under the peculiar conditions of mining, leads in another way to a somewhat unique relation between miner and foreman. For the miner's chance to earn money by loading coal depends on many other things beside his own workmanship and his own energy. Even on the little more than two hundred days a year that the average miner has the opportunity to work at all, his output is at least as apt to be limited by unfavorable physical conditions of place and seam and by unfavorable company conditions of car supply and the like as by any shortcomings of his own. A great section of the roof may have fallen in; the "bottom" may have lifted and upset the track; the place may have filled with water; a band of dirt in the coal may suddenly have doubled in thickness; or the seam itself may have vanished entirely in a fault. In each case the miner has unexpected "dead work" to do, and he must either lose the time it takes or get the foreman to make additional payment for these conditions. Or else he may on this account ask to be assigned to another place altogether, for "the energetic among the miners are always wandering about watching where the best places are."* But it is not only in their physical characteristics that places differ, and if a man asks for a transfer it is as often to get onto a heading where he thinks the cars are "running better" as to get into cleaner coal or better top.† For the miner's chance to load coal depends on his chance to get cars to load it into, and it is

---

*Hugh Archbald, *The Four-Hour Day in Coal,* p. 49. This book contains far the best description of coal mining methods from the viewpoint of the problem of supervision.
†This would not be true where the "square turn" is strictly enforced. See below, p. 59.

probable that this is the most important of all the factors limiting his output. Waiting for cars is a regular part of the work cycle of almost every miner and a highly characteristic feature of life in the mines. The extent of it is a matter of conjecture and controversy, with the opinions of operators and miners, and even of operators and their own foremen, at wide variance; but the fact of the waiting is a commonplace to anyone who has spent any time underground. The time studies cited above were taken largely with the object of determining in the particular case the time wasted in this way and in waiting for timber and other supplies. They showed that the total time lost (including lunch) was in 1911 over twenty-two per cent of the total time spent in the place and in 1912 just under twenty per cent; and even in the latter year about an hour and a quarter a day was spent in waiting for cars alone. But there is good reason for believing that this represents more regular work than the performance of the industry as a whole, and Mr. Hugh Archbald estimates the average waiting and loafing time not at twenty but at more than forty per cent.* Out of all this there are naturally many grievances piled up against the time of the boss's visit. It is true that the miners are not enthusiastic efficiency experts and that, as

*Ibid, p. 79.

Miss Evelyn Preston points out,* they are not as insistent upon one hundred per cent work opportunity as their engineering friends would have them. But cars obviously mean money— "I need nine a day," said one loader, "one for me, one for the old woman, one for each of the kids"—and just as the greeting "Good Morning" fitted the economic life of our farming ancestors, and as "Low Prices" has been suggested as its proper equivalent for modern housewives on their way to market, so the question, "How are the cars running?" has become a frequent and significant how-goes-it in the mines.

As a result of all this, the miner may be quite as anxious to see the foreman when he makes his rounds as the foreman can possibly be to see him. It is this fact that is missed by the hostile critic of the industry who writes that "the possibility of surprise tests does not exist."† The statement is perfectly true, but the impossibility is psychological and not physical. It rests not on any imaginary difficulty of "echoing footfalls" that would prevent the foreman from surprising the miner, but on the fact that he would see no point in stealing up on a man who felt himself an independent contractor with claims and grievances of his own fully as much as a

*In an unpublished thesis on "The Pit Committee" presented at the University of Wisconsin.
†Haring, *op. cit.*

workman subject to ordinary discipline. In fact I have heard bosses explaining failures in car supply and the like rather apologetically to the miners fully as often as I have heard miners apologetic to the bosses about *their* work; and the foreman is more the boss of the day men in the service of the miner than he is boss in the ordinary sense over the men at the face, and more of a safety inspector than either. "All the foreman needs to do is to get the cars to him," said one prominent coal man. That is overstatement, of course; but in a progressive company, "Keep the cars behind 'em all the time," or the prettier phrase, "Prompt service to the men at the face," which the Coal Commission delights to quote,* is likely to be a chief slogan for the foremen. And the emphasis on their safety duties is even more conspicuous, both in the mining law and the qualifications for foremen's certificates and also in the actual practice of the industry. Bonuses to foremen, in the very few cases in which they are given, are usually based on lists of "points" in which little weight is given to matters with which the men at the face have anything to do, and no weight at all to the amount of their output, or else they are given entirely on grounds of safety. "You

*U. S. Coal Commission, *Labor Relations in the Bituminous Industry*. By J. H. Willits and Associates. P. 24. (This document will be cited as the "Labor Relations Report.")

can't drive men in the mines," is a saying of the industry. "All you can do is take care of them."

This does not mean that the foreman has nothing to tell the miner, and no orders to give, when he makes his rounds. In many regions the enforcing of safety alone means a vigorous and continuous effort to compel the men to take down loose rock, instead of risking their lives by working under it, and to set their props and take proper care of their explosives, although there are other fields in which the "practical miners" stoutly insist that it is none of the boss's business where they set their timbers. The foreman may also see to it that the men are loading clean coal, though the chief inspection takes place at the surface and the standard is enforced by docking, and also that slate and rock are properly "gobbed" or piled out of the way. Very much less often he gives instruction in other matters of workmanship. A man's skill in shooting, for example, is of importance to the company as well as to the miner; too heavy charges mean too much slack and too little lump coal; but where directions in shooting are given they are much more likely to take the form of suggestions than of orders. The man who is to give them "has to know how to put his hand on a man's shoulder and say, 'Bill, I wouldn't do it that way. You may be right but let's see.' " The typical miner's attitude, moreover, is expressed in the indignant question:—"Why should the boss come dictatin' to me where to put my shots when I pay for the powder?" and most operators make no attempt at such "dictation." The president of one of their associations has even declared that he never heard of any attempt to supervise the placing of shots, and the ordinary foreman gives the miner little or no training either in this or in his other duties.*   Under the mining law of Illinois, which requires that a man must have a certificate of two years' experience at the face before he can be left alone as a "practical miner," it is a fellow miner and not the boss who swears to an affidavit to take care of him and teach him his trade; and throughout the industry —with or without such formal provisions—the miner does in practice pick up his trade almost entirely from his buddy or his neighbors and hardly at all from his boss.

Something of the same informality prevails also in the bargaining over dead work. The list of things paid for differs widely from region to region, and between union and non-union fields, and so also does the number of conditions for which payment is standardized and subject to little dispute; but in almost all cases there are

*See the discussion in *Explosives Engineer*, Oct., 1923.

many things left to the judgment and the hag-
gling of the boss and the individual workman. As
Mr. Archbald shows in his excellent analysis of
the foreman's problem, "he may be called upon,"
for example, "to guess how much should be paid
for cleaning up a fall of rock although he never
saw the fall and can only gauge his guess by looking up at the place from which it fell" and although he can do that "only after it has been
covered with timbering,"* and Mr. Blankenhorn
points out that in such conditions the men can
sometimes "capitalize their knowledge, greater
than that of the boss" and perhaps even "bargain a day's pay for two hours' work."† On
the other hand, there are no doubt many cases,
particularly in non-union mines, where the foreman is in a position to pay "almost anything you
want to give them"—as one boss told me—and
where a settlement unfavorable to the miner is
clinched with an "If-you-don't-like-it-you-can-
take-your-tools-out" from the boss. But usually the actual bargaining goes on in a manner of
considerable equality. Perhaps the loader has a
few feet of bottom to take up and claims a day's
pay for the job. "Well, I haven't been working
for some time myself," the foreman is likely to
say, "but if I couldn't do it in two hours I'd pay

*Archbald, *op. cit.*, p. 51.
†Blankenhorn, *op. cit.*, p. 1007.

you ten dollars." And so the argument continues until the bargain is struck. From one
non-union mine comes the story of an even more
informal debate. A miner was demanding extra
pay for working in a wet place:

"Hell, there's not much water over there."

"God damn you, there's six feet."

"It's not more than a ——— two feet."

The argument ended when the miner threw
his boss into the water and proved his point—
and kept his job! The story is not typical, of
course, and it may even be apocryphal; but it is
true that coal mining is the kind of industry in
which that sort of relationship is possible and
that sort of story believable.

A simpler explanation of the indiscipline of
the mine lies in the very infrequency and brevity
of the boss's visits. In most states the mining
laws call for a visit by the mine foreman or his
assistant once a day to each working place. This
is intended as a minimum, but it is no secret
about the mines that the number of visits is quite
as likely to be below as above the standard set;
for paradoxically the number of bosses in the
mines is as much less than that of other industries as the difficulties of supervision seem to be
greater. One hundred men to a boss is a common ratio inside the mines. In most trades that
would be considered ridiculously little super-

vision even for gang work above ground, and to expect one man to give any sort of detailed guidance to the work of a hundred men as he goes the rounds of the isolated working places in the darkness of the mine is a little like asking the postman to oversee the housework in all the houses on his route. "Five miles will be a short day's tramp for him," writes Mr. Archbald. "A foreman who has visited only thirty places in the four hours of the morning will consider that he has done an easy day's work. Four hours—two hundred and forty minutes, thirty places—eight minutes to a place including the coming and going." And the coming and going itself is no small matter. "The distance between any pair of men is likely to be at least one hundred and fifty feet. In passing from one room to the next, a man will have to duck his head as he goes through the cross-cut, perhaps crawl on his hands and knees when the coal is low."* It is no wonder, then, that if a boss meets a miner in the heading and asks, "Is your buddy working today?" he is likely not to go up to the face at all, and that the visit is often skipped altogether. But even this degree of supervision is much greater than that in some important mining regions. The Illinois law, for example, requires a visit only once *every two weeks*, and there are

*Archbald, *op. cit.*, pp. 39, 49.

mines in the state in which a boss rarely enters a place except to direct the placing of a cross-cut or to bargain over dead-work. The chief reason that the miner is a remarkably unbossed workman is that the foreman is so rarely on hand to boss him.

The miner is an isolated piece worker, on a rough sort of craft work, who sees his boss less often than once a day. It is on these facts and their corollaries that much of the miner's freedom or the mines' indiscipline depends. But its most conspicuous single expression is in the miner's privilege of coming to work and leaving his work as he chooses. The starting time is now fairly regular, to be sure, although in non-union mines men sometimes go in early to "move some slate while they are feeling stout." But the miner's right to go home when he pleases is still his throughout almost the entire industry. Usually he does not even have to check in and out, and if the mine is a drift or a slope he simply walks out the level heading or up the incline into the open air. If it is a shaft mine, it is not quite so easy; he needs a cage to take him up, and the cages are busy hoisting coal instead of men, but in most cases the man finds it possible to get out of the mine without very great delay. "By quitting time in some towns, most of the miners will be out of the mine and many of them will be washed up and down town again;" and it has even been estimated, on the basis of the scant data available, that the miner spends only about six hours a day in his place.* It is true that a few companies make vigorous efforts to hold their men in the mines the full eight hours, but even there it is much more often a matter of persuasion than of compulsion. Sometimes, in fact, it seems to be nearer coaxing than persuasion, as in the case of the foreman who explained that what he did when he met a man coming out early was to say, "Wouldn't you load a couple more cars if I got the motorman to give them to you right away?" The remark itself goes far to explain the survival of the custom. I do not mean that waiting for cars and no work to do are the only things that send the miner home early. He goes out, just as he stays home from work altogether, for all sorts of reasons; he may quit to butcher a calf, for example, or to help cook the dinner for the boarders on the night shift, or simply because he "knows when he is tired," or because he has earned what seems to him enough for the day. And there is much truth in the common belief that, either on account of the real strenuousness of their job or as a matter of long tradition, miners are not good for very much heavy work after lunch. But

*Ibid, pp. 63, 79.

the cool dampness of the mine makes it a poor place to loaf after the sweat of picking and loading; the fact that so often no work is ready for them is a major reason why miners actually quit early; and it is certainly a chief reason why the tradition of the irregular day has grown up, and a chief reason also—since it is the common sense of the industry that there is no use in keeping men in the mines when there is nothing for them to do—why there is no very widespread attempt to break up the practice. This "untrammeled privilege," then, to which operators sometimes point, has its roots in the slacknesses from which the miners suffer as well as in the felt freedom of the miners' position as piece workers in places of their own. But it is no wonder that outsiders should consider it the most astonishing fact about the organization of the industry, or that it should be cited as the chief proof of the miner's freedom. "He comes to work when he pleases and he comes out when he pleases," said one operator. "We are not able to control him."* The industry in which even convict laborers could not long ago choose their own quitting time† is very far—for better or worse—from the regularity and the regimentation of the time clock and the factory whistle.

*District 2, *Joint Conference Proceedings*, 1903, p. 247.
†Bituminous Coal Commission, *op. cit.*, Vol. VIII, p. 549.

The Machine Runners

It is in the working life of the men at the face that the quality of indiscipline appears most sharply, but many of its elements extend also to the small group of employees, four or five per cent of the total number, who operate the mining machine. These men also are piece workers, at least in the majority of cases, and with the men at the face they share the term "miner"—which by the strict usage of the industry does not apply to other mine workers—and much of the tradition of indiscipline that goes with it. On the other hand their actual work is totally unlike that of the pick miners or loaders, and out of it comes a quite different type of individual responsibility and a different type of independence.

The machine runners work in teams of two, the cutter and his scraper or helper, moving their machine from place to place and cutting the coal for a number of loaders. This means of course a very different position in the organization of the mines from that of the loaders themselves or of the pick miners. A pair of men working in a place of their own have little to answer for to anyone but themselves, but if a single machine crew lays off, a dozen loaders may have to go home for lack of work; the job carries with it a definite responsibility for the work of a number of places. As a result the machine runners are held more strictly to account for failures in attendance and workmanship than the men at the face, and they do come to work much more regularly and on the whole put in longer hours. But for all this they are surprisingly little supervised; in the course of their actual work they are if anything less bossed than the other miners. Oftentimes they do their cutting at night, when the boss for whose section they are working is not even in the mine, and in any case they do a large part of the planning of their own work. The width and placing of the cuts are standardized; the boss may have marked the places to be cut, or if he happens to meet the crew on his rounds he may tell them that "that Hungarian fellow down there on seventh right needs a cut right away;" but that is about all the supervision comes to. Very often the machine men are simply required, as one union contract puts it, "to keep the cutting up" on their own responsibility in a given section of the mine.*

The machine runners, then, go from place to place largely according to their own judgment; and the job itself is one in which they can feel some pride and some sense of control over their

*District 17 (West Virginia) *Monongahela Agreement*, 1923, Sec. 10.

machine.   The machine they use looks, at least
to a keen eye on the *Survey Graphic,* "like a sea
turtle with a sword-fish's snout, spinning chains
in and out of itself like a spider."* The snout
is the cutter-bar, which projects six feet or so in
front of the machine, and around the edge of
which the cutting chain is attached.   "On starting the machine the endless chain with the steel
cutting bits revolves"—this is from the Coal
Commission's more sedate account of the "spinning" process—"and is automatically forced forward against the coal."† These bits do the
actual cutting as the bar is forced its full length
into the coal and then across the face.   As it
moves across, the cutter himself merely guides
it—I have even heard old cutters boast that they
could fall asleep in the process and wake instant-
ly when the bar hit a sulphur-ball or other ob-
struction—while the scraper clears away the bug
dust that is made as the steel bits grind up the
coal.   This is the core of the job; but the hard-
est and most active work and the greater part of
the skill come in setting the "amazing brute of a
machine" in place to begin the cut,—moving it
off the truck on which it travels, skidding it over
to one corner of the face and setting up the two
jacks to which it is anchored—and in starting

*Survey, Vol. XVII, p. 991.
†*Thompson Report,* p. 4.

THE LOADING MACHINE

Courtesy of Joy Machine Co., Franklin, Pa.

and stopping it accurately at the right place in the seam.   The job is not an extremely delicate one—no jobs in the mines are that—and it does not require very great mechanical knowledge; but it does call for deft and well-timed and very strenuous movements performed at top speed.

Both in pay and in pride of work the machine runners are the aristocrats of the industry. Their life has less of the sheer indiscipline of the mines than that of the men at the face, but their greater responsibility is a highly individualized one, and they work in almost as complete independence.

THE COMPANY MEN

The language and the traditional attitude of the mines make a sharp distinction between the men who work by the day and the men whose jobs have already been described. The day men are "company men," and a prominent operator remarks that they take their jobs "with the understanding that they are only assistants to the management."* It is a phrase that he would not think of applying to the tonnage man; they "are miners," as another operator put it, "and therefore not in our control."† And even

*Bituminous Coal Commission, *op. cit.*, Vol. VI, p. 304.
†United States Industrial Commission, *Proceedings*, Vol. XII, p. 82.

the company's right to "direct the working force," which is affirmed by the union agreements, is sometimes held not to apply to the miners; "by working force," says one contract, "is meant those men for whom a daily or a monthly wage is agreed to."* The miners are self-directing—so the theory seems to run—while the company men are a disciplined working force directed by the company; but the facts of actual practice by no means divide so neatly. It is true that the day men do not share with the men at the face the special claims and the special psychology of the independent contractor, and they do not go in and out as they please. The day man is a worker whose time is kept in some boss's book. And it is true also that the company men are held to somewhat more regular attendance and are shifted from job to job somewhat more freely than the miners. But in actual fact the miner is somewhat more bossed, and the company man often very much less bossed, than the theory would suggest; and even for day men there is little in the organization of the industry that approaches the regularity and discipline of a scientific management factory.

The actual supervision under which the company men work varies widely from job to job according to the ease with which the foreman

*District 20 (Michigan), *Agreement*, Sec. 1.

can see them at work. The outside day men, the ten or fifteen per cent of the employees who work on the surface, are naturally the easiest to boss. The largest group of them, the tipple gang, who do the work connected with the dumping of the coal from the mine cars into railroad cars, are much like gangs of men in other industries that have straw bosses of their own and that are often in plain view of higher officials. The hoisting engineers, in shaft mines, "are good sober men who will work," as one operator said, "any time we want them."*   The carpenters, blacksmiths, machinists and the like work very much as their fellowcraftsmen do in other industries, although even on the outside there is often much informality.   "I've been here fifteen years," said one boss carpenter, for example, "and this is the first time anybody ever told me not to let men hang around here and talk."†

Inside the mines there are many company men whose work is harder to supervise, at least by seeing, than that of the men at the face themselves. The miner has a definite working place where the boss can expect to find him, but that is true of very few of the inside day men.   The men who do the easy and monotonous task of minding the pumps, and the bottom men who

*District 2, *Joint Conference Proceedings,* 1904, p. 239.
†Archbald, "Mister Super," *Survey,* Vol. XVII, p. 1022.

shift the cars on and off the cage in a shaft mine, work in definite places and places easily accessible to the foreman; but almost all the other company men inside, except the trapper-boys who open and close doors at scattered posts throughout the mine, have traveling or roving jobs. At a given time a motor crew, for example, might be anywhere along several miles of track; the pairs of maintenance men would most of them be even harder to find. Here again "the possibility of constant supervision does not exist;" and since in this case the company has more to lose, the work of these day men raises even more acutely than that of the miners the special riddle of the industry: *how can you boss men that you can't see?* What *is* the method of supervising these men scattered so widely in the darkness of the mine? "Pretty good system to it," answered a former miner. "A sort of knocking." He went on to explain that the motorman gets "knocked" if the coal does not come out fast enough, and that he in turn "knocks" the tracklayer if a bad rail has wrecked his cars. Or at the other end the miner gets the boss to "knock" the driver if the cars are slow or the tracklayer if a new switch is not laid. And the foreman finds other things to "knock"—wires down, timber badly set, and the like—as he makes his rounds. Supervision by knocking is in fact a shrewd and accurate summary of the methods of bossing in the mines.

This means of course that in the first instance —before the knocking——a considerable individual responsibility falls upon these workers. That is very clear in the case of the largest group of them, the motormen and their trip-riders, or brakemen, and the drivers. Each mine has a complicated system of transportation underground in which all the many terminals from which the coal is gathered are "continually advancing,"* as the Coal Commission points out; yet it is almost always a railway system without schedules and without train dispatchers. The control of haulage is one of the chief of the foreman's many problems, and sometimes there are motor bosses or driver bosses with a certain degree of authority; but in practice the problem is very largely "handled at the will of the man driving a mule or a locomotive."* Each main line motorman has a certain territory to haul the trips or trains from; each gathering motorman or driver has a smaller section in which he is responsible for the gathering and distribution of the single cars; and most of the actual arrangements and decisions involved are made by the men themselves. "Handling the trips and signals and what headings to go into and keeping

*Thompson Report, pp. 15-16.

the turn and all that sort of stuff"—as a union leader defined the responsibility*—is in the hands of the main line motor crew. Keeping the turn, or distribution, between the individual working places in a section, is the job of the driver or the crew of the gathering locomotive. A proof of this lies in the fact that the union in its efforts to secure an equal distribution of cars has sometimes brought pressure to bear directly on the drivers instead of on the company, and in the suggestion that it is often advisable for the man in the place to win the driver's favor by his wife's skill in pie-making. And the extent of the motorman's discretion in another direction, that of the control of his machine, is suggested by a foreman's account of a particular motorman's indiscretions with his:—"He was reckless both in running and plugging the motor (stopping by reversing), and jerked the cars when he was down in 'D' headings. He would try to get out with a bigger load than he could. He would back it up and wreck cars. I told him to split it and bring it out twice but he didn't do it."*   The case is not typical; the man was fired —though not until the electrician had "thousands of times coaxed him to quit it;" but it does indicate the variety and degree of responsibility that go with the job.

*District 2, *Arbitration Decisions,* Sept. 4, 1919.

The motorman is perhaps the most skilled of the company men inside; in the cases where the cutter is paid by the day the rates for the two jobs are usually the same; but the skilled maintenance men,—the tracklayers, the timbermen, the bratticemen who build the canvas and wood and concrete walls and doors and partings for the elaborate and constantly changing system of ventilation, and the wiremen and pipemen—work under conditions of very similar independence. Sometimes they also—particularly the tracklayers—have definite sections of the mine assigned to them in which they are supposed to find their own jobs and in which they are responsible for the maintenance; but more often they are sent from one part of the mine to another at the boss's orders. They usually work in pairs, however—wireman and wireman's helper, and so on; and their jobs scatter them so widely in the intricate network of the mine that once they are at work they are as a rule less often seen by the boss than either the transportation men or the men at the face.

This does not mean that the work of either of these groups of day men is entirely uncontrolled. In the first place they are selected somewhat more carefully than most of the other employees; although here again they learn their jobs almost entirely from each other—brakeman from motorman, tracklayer's helper from track-layer, and so on; and although the interview that begins when a new worker slouches into the foreman's shanty with the question, "Settin' on any motormen to-day?" is likely to be informal in the extreme. In the second place, the fore-man has a general idea how large a section a motor crew or a tracklayer can take care of, or how long a bratticeman and his helper should spend in walling up a break-through; and, al-though the actual work of the day men is rarely recorded and rarely measured, and although it is easy for the efficiency experts to show that in these matters the common phrase of the mines, "the foreman knows," is far from literally true, nevertheless the system of knocking has some de-gree of force. Careless timbering is easily seen, reckless driving and badly-laid switches an-nounce themselves in wrecks, and too much loaf-ing is likely to send a man back to loading coal where he has to "work or starve." These men are somewhat more subject to discipline than the miners, but it is clear that their actual work is done under conditions not of constant bossing but of remarkable independence. There are also common labor jobs inside the mines, such as that of the road cleaners; but the skilled main-tenance men together with the transportation men make up well over half of the inside day force, and the characteristic company job underground is one that calls for some degree of skill and a high degree of self-direction.

These semi-independent pairs of inside day men, the men at the face, and the machine runners—these three groups whose separate types of independence and indiscipline have been described—make up three-fourths of all the mine workers; and much of the tradition of the miner's freedom extends beyond them to the other employees, inside and outside. Coal mining is an industry in which the majority of men are piece workers under very slight supervision; it is also an industry in which a great number of the other employees—though this fact is less often recognized—are on jobs that are quite as unstandardized and almost as little supervised as those of the miners themselves.

CHAPTER III

THE RIGHTS OF THE MINE WORKERS

THE technique of coal mining leads very naturally, as has been shown, to conditions of indiscipline, but it does not completely explain them. For room-and-pillar mining may in fact be carried on with thoroughly submissive workmen,—with serf labor as in the early days of British mining or with negroes still under the spell of slavery as in isolated Southern mines to-day; and the prevailing freedom of the miners is in part a matter of vigorous human tradition.

The teaching of this tradition goes on largely in the working place itself. A certain anthracite miner, for example, was telling how he had given a lesson in the ways of the mines to the newly-landed "hunky" who was working as his laborer. "Come here, Frank, says I. Here's the boss. Don't work. Always sit down when the boss is around." The story comes from an allied industry, and the particular custom in question seems within the last generation to have dropped out of the bituminous mines, but there is no doubt that the process of teaching is typical of both industries. Just as the younger worker picks up from the older one the greater part of

THE MAN-TRIP

Photo by Lewis W. Hine.

Reproduced by permission of The Survey-Graphic.

his craft knowledge, so also he learns from him most of his attitude toward the boss and toward the industry.   The vivid phrase of a union leader, moreover, throws light on another part of the process.   "It's the *gob pile oration*," he said, "that makes the miners less submissive than other workers.   A miner hears the boss bawling out the man in the next place.   Pretty soon there's a delay or something and the men get together on the piles of slate out in the heading and talk it over."   "The bolder ones," he added, "teach the others;" and out of it comes a common attitude toward the boss.   "It's the delays," said another union man, "that make the miners debaters."   It is certainly the delays that bring them together in the small groups on the gob piles, and very likely it is true that the "orations" that are delivered there are of more importance in setting the manner of life in the mines than all the speeches of the official union leaders outside.   There are many other things also in the daily life of the miners that make for their solidarity.   There is a suggestion of this in the group of men gathered at the end of the day at a crossroads inside the mine waiting to be jostled out together in the "man-trip,"* but it is in the villages to which the men go home at night that the background of common feeling is

*See illustration opposite p. 57.

most evident. "We haven't got a farmer or anybody else in the town," said a miner, "except one hotel-keeper."* That is the extreme, but the mining of bituminous coal is for the most part carried on far from cities and far from other industries; and the miners, more often than almost any other workers, live in isolated camps and villages of their own. The communities, to be sure, are usually made up of a great mixture of nationalities; but they are communities in which the talk in the pool room or on the log across from the company store is largely the talk of the one industry,—communities, most of them, to which all the men come home black.†

The talk of the working place, the orations of the gob pile, the public opinion of the mining village,—there are thus many ways, even aside from the formal organization of the union, in which the traditions of the miners are passed on and strengthened; and, as a matter of fact, some of them are older than the union itself and many of them are forces to be reckoned with even beyond the boundaries of the union fields. John Mitchell and other leaders of the United Mine Workers have referred to the importance of the "conditions that have grown up with the indus-

*District 2, *Convention Proceedings*, 1908, p. 21.
†This same fact has been noted as an explanation of the unusual solidarity of the British miners. G. D. H. Cole, *Labour in the Coal-Mining Industry*, pp. 6–7.

try"* and "the practice of the trade"† that is
older than the union organization; and the tradition of the "square turn," for example, "that no
man should have more cars than another man,"
which is often thought of as merely a union restriction, is clearly of much longer standing.
"Even without contracts," said a prominent operator in the early years of the union agreement
in Central Pennsylvania, "the question of un-
equal distribution has always arisen. That is an
old question that has been up for twenty-five
or thirty years."‡ And there is even reason for
believing that this and other "customs and conditions" of the industry may have had a still
longer history in the British mines where so
many of the early American miners learned
their trade.

Coal mining is an old industry with old traditions as well as in part a union industry with
union traditions, and many of the characteristic
elements of this old indiscipline may be found in
the attitudes of unorganized as well as organized
workers throughout the greater part of the industry. "Even the non-union miner," declares
an active member of the U. M. W., "is freer
than a factory worker." There *are* mines, to

*Bituminous Coal Commission, *op. cit.*, Vol. VI, p. 296.
†Industrial Relations Commission (1916), *Final Report and Tes-
timony*, Vol. I, p. 415.
‡District 2, *Joint Conference Proceedings*, 1905, p. 101.

be sure, in which it can be said of the workers that they "only know what the coal company teaches them" because they are fresh from the farms of Virginia or Poland or because "the majority of the men is just the men the company made them out of the timber that lived here when a mine was an unheard-of thing;"* but the ordinary non-union mine has among its employees a number of old miners and often a group of former union miners, and from these men the rest of the workers acquire much of the traditional attitude of personal independence. The miner's right to go home when he pleases does not end at the union border; in fact, one miner told me that he "had to come from West Virginia to solidly organized Illinois to find a mine where they'd discharge a man for coming out early." Absenteeism is no less frequent in the non-union fields,† and it was a non-union foreman who declared that "if a man has been in here loading three or four days, he's done something, and if he lays off the next day, that's his own business." Even in some of the fields most securely beyond the reach of organizers, the customary rights of the mine workers are felt as a serious obstacle to the stiffening of

*From unpublished material collected for the U. S. Coal Commission.
†U. S. Coal Commission, *Irregularity of Employment in the Coal Industry*. By Horace B. Drury.

discipline, and I have heard non-union miners stating as eloquently as any union ones the advantages of mining as a job in which you were "more your own boss" than in a factory.

The democracies of the gob piles assert much of the traditional freedom of the miners even beyond the union frontier, but there is no doubt that on the whole they speak much more boldly in the union mines. "Non-Union Towns are Towns of Fear," declared a card circulated during the 1922 strike:—*

"Non-Union Miners Have Lived in Fear!
    Fear of the Boss
    Fear of Spies and Spotters
    Fear of Gunmen and Coal and Iron Police
    Fear of anti-union Civil Authorities
    Fear of the Blacklist
    Fear of Evictions."

To be sure, the communities in question, even those in which all the weapons named are used quite ruthlessly against organizers and strikers in time of struggle, do not usually seem like "Towns of Fear" to their loyal inhabitants.† Sometimes, even, they may seem like Kingdoms of Love, at least according to the ingenuous

*By District 2. Reprinted in full in Heber Blankenhorn, *The Strike for Union.* See also U. S. Coal Commission, *Report on Civil Liberties;* Winthrop D. Lane, *Civil War in West Virginia;* and numerous magazine articles.
†This point is well brought out by A. F. Hinrichs. *The United Mine Workers and the Non-Union Fields,* p. 11.

language of a poster issued by a certain church
of colored miners:—

> "8 p. m. The Hon. Mr. ———, the owner
> of this beautiful plant and the greatest
> Negro friend in West Virginia, will address
> the audience. 8.30 p. m. Mr. ———, *the
> second man in the Kingdom*, the honored
> and highly respected *superintendent of this
> town*, the man we hope whom when the
> mantle falls from Elijah it will fall upon
> him as it did upon Elisha, will also address
> us." (italics mine.)

But in either case, whether the motive be fear
or a sort of plantation loyalty, it would be surprising if the miners in these villages were
very vigorous in asserting the traditions of the
workers at the points at which they ran counter
to the wishes of the management. And surely
there must be some reflection underground of
the difference between the community in which
the coal company which owns the land is all the
local government there is, as is frequently the
case in the non-union fields, and the community
in which, as sometimes happens on the other
side of the line, the president of the local union
is elected mayor or burgess. To the question,
"What do the men think of a certain change?"
two non-union superintendents gave the significant answers:—"What does that matter?" and

"We have no organization here." No union boss could have answered so nonchalantly. The organization would have furnished a channel through which the men's opinion would have made itself felt, and in case after case the union appears as the great bulwark of the traditional independence of the miners.

The relationship of the union to these individualistic rights, however, is not always one of direct defense. In fact there are certain elements in the old indiscipline that the organization expressly disavows. It makes no claim for the right of miners to go home early or to stay away from work at will. "And if they do lay off," declared one leader, "as some of them do by getting too much booze . . . they will have to settle their own grievances with the employer . . . He is at liberty to discharge them if he sees fit."* This position, moreover, is now taken in the majority of agreements. "Should any employee absent himself from work for two days," reads a typical provision, "or persist in working irregularly, unless through sickness, or by first having notified his Foreman and obtained his consent, it shall be construed as a dischargeable defense;"† and there are a number of districts also in which the contracts either

*District 2, *Joint Conference Proceedings,* 1908, p. 35.
†District 17, *Monongahela Agreement,* 1923, Sec. 37.

"jointly recommend and request" or else require that the man remain in the working places "the full eight hours, or such part of the eight as they have work to perform."* But for the most part these rules are "treated with impunity by the employees," and no doubt a sweeping campaign of enforcement would meet with more opposition, in spite of the agreements, in the union than in the non-union mines. A passage from the diary of Mr. Edward A. Wieck, a mine committeeman in Southern Illinois,† throws a much clearer light on the actual relation between the union and the old custom of irregularity. The mine manager had just posted notices requiring the miners to stay in their places until 3:45 in the afternoon. "Such orders are strictly according to the contract," comments the diary, "and have been in effect in various parts of the district for twelve or fifteen years, but . . . no doubt some of the men will buck this, which will mean work for the committee to do trying to get their jobs back." The boss asked the committeemen to take an active part in getting the miners to obey the rule; they were not quite willing to put themselves in the position of seeming to "help the management impose onerous conditions on the men" but decided instead to "ignore

*District 17, *Northern West Virginia Agreement*, 1923, Sec. 4.
†Portions of this manuscript were printed in the *Atlantic Monthly* for July, 1924, under the title, "A Coal Miner's Journal."

the matter and let the boss enforce it as he saw fit." "I called his attention, however," continues the committeeman's account of the interview, "to the fact that it was a new departure in this field to insist that piece workers work up to the last minute. 'Customs and habits of a lifetime are not easily changed,' I reminded him. 'You can't expect to put up a notice and get immediate results, especially when you are dealing with men.'" And the final comment in the diary is:—"I think there are no discharges imminent." The picture is doubtless thoroughly typical. When a movement toward the new discipline starts, "the men don't like it," and the leaders of the union are more interested in protecting the jobs of their members than in campaigning for efficiency; but it is clear that the basis of the resistance lies in the old customs of the mine workers and not in anything the union has added to the tradition.

The union fights neither for nor against the position that a man "ought to know when he is tired" and when he feels like working, but it does assert and defend many of the other customary privileges of the miners. A number of these rights are written into the joint agreements, not merely under the head of the significant clause —"customs and conditions to remain the same" —but in specific provisions as well. The principle of the square turn, or equal distribution of mine cars, for example, that expresses the traditional feeling of the miners that "they ought to get a fair share of the work that is going,"* has become institutionalized in the contracts in an elaborate body of rules and exceptions and safeguards; and the right of miners to separate and special working places is either stated or implied in all the agreements. In most of the districts there are clauses making the arrangement definite,—providing, for example, that each man shall have a separate place, or that a pair of loaders shall have two rooms unless the machine runners can keep them in steady work with one, or that two men shall have three rooms, and so on. Even in the longwall mines of Texas and Northern Illinois,† the contracts expressly provide that the miners shall work by ones and twos and "assume the responsibility of the place"‡ very much as in the more usual type of mining. There are occasional rules also, probably reinforcing a more widespread custom, that declare the right of the miner under certain conditions to choose his own "buddy" to share the place with him. Moreover, the union is almost everywhere successful in its contention that the

*District 2, *Joint Conference Proceedings*, 1905, p. 99.
†In longwall mining the working face is an unbroken line along which many men work. See above, p. 20.
‡District 12 (Illinois), Subdistrict 1, *Agreement* Clause 8.

foreman "cannot force a miner to leave his place and go and drive mules (or do other company work, unless he is willing to do so, and that he cannot stop the cars on such a person" if he declines to make the change.* Under certain circumstances, to be sure, one man's place may be given temporarily to another worker, but the rule expressly provides that "such loader must leave rooms in as good condition as found."† One agreement provides that a man's place must be held for him during his absence on union business or during an annual vacation of two weeks, if he notifies the company in advance;‡ and the actual practice of the industry, as indicated in the decisions interpreting the agreements, goes much farther than this in confirming a miner in his property right to a particular place in the coal. A number of such decisions hold that a man has a claim to his old place even after a period in which the mine has been shut down, and another case throws an even more interesting light both on the customs of the industry and on the position taken by the union. A certain miner in Central Pennsylvania was slightly injured in the mine and stayed away from work

*District 2, *Convention Proceedings*, 1906, p. 13.
†District 2, *Agreement* 1920, Rule 8.
‡For this and a number of other rules cited in this chapter, I am under obligation to Mr. A. E. Suffern who has prepared a careful digest of the scale agreements for the Institute of Economics.

for some time without sending word to the boss.
"The mine foreman held the place for him for
three weeks, and then upon learning that he was
working about his store, driving a team in de-
livering goods, etc., gave his place to another
miner." At the end of *six* weeks the man re-
turned to the mine and demanded not only re-
employment but re-employment in his own old
place. The union backed his claim; and the
case was referred, according to the usual pro-
cedure, to the commissioners, or representatives,
of the union and of the operators' association.
"In view of . . the fact that Gravich followed
the advice of the physician, but did not keep the
foreman advised," reads their decision, "and his
place having been given away to another miner,
we decide that Gravich be reinstated, either in
his own place or in an average place in the
mine."*

In defending claims like this and in enforcing
these provisions in the contracts, the union appears as the supporter of certain elements in
the old indiscipline; but its most important con-
tribution to the individualistic freedom of the
mine workers comes in the protection it gives
them from arbitrary discharge. The contract
provisions on the subject, to be sure, always
start with the statement that the union has noth-

*District 2, *Arbitration Decisions*, 1915, Bulletin No. 7, pp. 1–2.

ing whatever to do with it. "The right to hire and discharge," the contracts declare, "is vested exclusively in the operator, and the United Mine Workers shall not abridge that right;"* but this ringing declaration is contradicted, and this right is certainly shared and abridged, both in the provisions of the contracts and in the practice of the industry. In the first place, the general power of discipline is limited in all the agreements by clauses that prescribe specific penalties for particular offenses. "A miner who sends out 'dirty coal' shall be suspended one day for the first offense," reads a typical contract, "three days for the second offense, five days or discharge for the third offense within any thirty days."† Moreover, the very section of the contract that states the principle of the employer's right to hire and fire almost always follows it with a "but" and an important qualifying clause. "It is not the intention of this provision," explains one such statement, "to encourage the discharge of employees . . on account of personal prejudice or activity in the United Mine Workers of America."‡ The more usual type of contract goes even farther and specifically recognizes the right of the union to

---

*District 2, *Agreement*, Rule 15.
†*Ibid*, Rule 8.
‡District 6 (Ohio), Subdistrict 1, *Agreement*, p. 178.

"take up" all cases of discharge. If "in the opinion of the mine committee, . . . any employee . . has been unjustly suspended or discharged, the mine committee shall endeavor to have him reinstated;" and if necessary the case shall be carried up through the various "courts" provided for the adjustment of disputes.* The employer's right to discharge is absolute, says the contract, but if he uses it "unjustly" the man must be put back to work and, in many cases, be given back pay for the time lost!

The curious logic of this position reflects the modification of the contract with the growing power and activity of the union, and even now its actual power outruns the contract provisions. "You must have a just cause"—and, as the Coal Commission points out, good evidence—"before you discharge a man" if you want the penalty to "stick;" and there are many operators who would amend the statement to read that you can't get rid of a man even if you *do* have a just cause, or at least that "you have to bring it around to the point where it would be absurd *not* to discharge a man."† Certainly it is often felt to be the duty of the mine committee to contest every case of discipline that arises. An Illinois committeeman, for example, was look-

*District 2, *Agreement*, Rule 15.
†*Labor Relations Report*, p. 34.

ing ruefully at the enormous pile of "clod" that a certain miner had tried to load out as coal. Another miner came up and grinned sympathetically, "You'll have a hell of a time getting that fellow's job back!"* Evidently where there is an active committee, no man need lose his job, no matter how flagrant the case, without a hearing and without benefit of economic counsel. The benefit of the doubt, moreover, is usually given to the miner, and very often the decisions of the commissioners or arbitrators take the form of sustaining the contention of the company but recommending that the man be put back to work! In certain districts, also, the general statement of the right to hire and fire has been so overshadowed by other provisions that it is officially held that discharge cannot take place except for offenses that are definitely stated in the agreement. It is said that in Northern West Virginia, for example, men who have been discharged for running disorderly houses in the mining camps have been put back to work on the ground that "the agreement prescribes no special kind of conduct outside working hours;" that men who have been fired for loafing have been reinstated on the ground that "the scale agreement does not say just how hard

*Wieck, "A Coal Miner's Journal," *Atlantic Monthly*, July, 1924, p. 8.

a man has to work;" and that men who have been discharged for swearing at their foremen and for "threatening them with bodily harm" have been given their jobs back because these things were not "dischargeable offenses under the agreement."* But these cases are exceptional, and the careful discussion of the matter in the Labor Relations Report concludes that "over most of the areas where relations under union conditions have obtained for a longer period, . . . discharge, when merited, can be made and the maintenance of effective standards insured at the same time that unfairness in discharge is avoided. . . . The outstanding fact (however) is that discharges and other disciplinary acts are very much more difficult than formerly" and very much more difficult than in the non-union mines.† It is not true that the union's power makes discharges impossible. It is not even proven that it is what makes them infrequent. Discharges are rare enough, to be sure, in the union fields to-day—one company with a payroll of six thousand fired only twelve men in a year in which it was taking advantage of a depression to rid itself of "reds;" but it is not clear that they were any more numerous in

*See the summary of such cases on p. 35 of the *Labor Relations Report*.
†Pp. 35, 33. The Report gives a number of cases of discharges that have been sustained in the most highly unionized fields.

the past—an operator of fifteen years ago declared that in his experience "suspensions (were) simply and solely for loading dirty coal;"* and it is not even clear that they are much more frequent in the non-union fields, except where the weapon of wholesale discharge is used as a campaign measure to get rid of union men and "agitators" or to force the men to sign the "yellow-dog" anti-union contracts.   Both union and non-union bosses know how to "freeze men out" or "ease them off" by giving them poor working places, but in ordinary operation outright discharge is comparatively rare in either part of the industry.

It is not the number of discharges, however, that is the important point for the study of freedom and indiscipline, and the essential and significant difference is that under union conditions discharge cannot take place as a simple act of the employer's will or the foreman's temper.   It is the effect of this on the attitudes of men and bosses underground that gives meaning to the organizing slogan, "Join the Union and Quit Being Afraid of the Boss!" and the assurance of a hearing and the sense of power given by the organization go far in making the miners readier and bolder in asserting their individualistic privileges.

*District 2, *Joint Conference Proceedings*, 1909, p. 14.

Sometimes, also, the very discharge cases themselves are fought by the union officials on the ground, not of the contract rights of the employees, but of their traditional—or "natural" —rights as mine workers. An arbitration board in Central Pennsylvania, for example, was hearing the case of a pair of machine runners who had been fired for refusing to obey the foreman's orders. "The men considered the place wasn't safe to cut, the foreman ordered the men to cut the place, and they refused to do it." "The posts were cracking all around," testified the cutter, "and in the morning a fall was down pretty close to where he wanted me to go in and the post was broke in two . . . I refused to go in there while the roof was working." "You see," explained the mine committeeman, "when they put up a jack to get in, the machine might loosen the roof up and maybe it is easy to come down." To an outsider this seems like a vivid picture of danger, but the more prosaic testimony of the company showed that the place was in fact cut that same night without accident by the foreman and another worker. "The question involved," summed up Mr. Mark for the union, "is whether a man is supposed or required by the contract or this Board or the law of the State to work in any place that he considers as dangerous to life and limb." "The whole matter is determined by law," argued the operators' commissioner. "The mine is exclusively in the hands of the mine foreman. If this Arbitration Board is going to lay down the principle that every man for himself shall determine for himself whether he will work in a given place or not, you are laying down a dangerous principle." "By Heavens," declared the union leader, "I would not let any mine superintendent or mine foreman determine for me whether my life was safe to work in a place or not;" and it was on this appeal to the logic of individual rights that the union won the decision for the men.*

It is in ways of this sort that the union intensifies the customary independence—or indiscipline—of the miners, and in the process it adds an important new element to the tradition. For in defending the miner's freedom in regard to his own work, the organization secures for the miners as a group something of a collective freedom in regard to the running of the mines. This new power, to be sure, has grown up without definite intention on either side. The United Mine Workers of America drives forward towards its major goals of wages and hours under strong conservative leadership and with

*District 2, *Arbitration Proceedings* and *Arbitration Decisions*, 1921.

little attention to modern doctrines of "workers' control." And surely the operators have not intended any such sharing of authority. "The management of the mine and the direction of the working force . . . are vested exclusively in the operator" according to the emphatic language of the contracts; and this provision is often reinforced by a clause that attempts to limit the activity of the mine committees to "the investigation and adjustment of disputes" and declares that they "shall have no other authority or exercise any control nor in any way interfere with the operation of the mine."* But the frequent disputes over these clauses and the frequent charges of "interference with management" are good evidence that the "dividing line" is very much blurred in the actual practice of the industry. This fact was well brought out in the course of an early scale conference in Central Pennsylvania. "We are coming to the point," said Mr. J. L. Spangler, "where we will have to determine just how far we have any authority as operators in the management of our own works and how far we concede that our men take part in management." The remark led to an honest attempt to clarify the issues in which Mr. W. B. Wilson, later Secretary of Labor, was the chief spokesman for the miners:

*District 2, *Agreement*, Rule 15 and Rule 4.

"I still don't see," persisted the operator, "how far you go in the management, the joint management of our mines."

"So far as our interests go, Colonel."

"Well, your interest, it seems to me, goes to the extent that if you get your wages and the proper conditions . . . under an annual scale that is made and we keep that contract with you, that is the limit. You ought to be satisfied. But when you ask us to submit to you any changes in the hiring and discharging of men, or in doing anything outside of that, you are interfering with our business."

Mr. Wilson met this with a partial disclaimer. "Our position . . . is this, that (in case of) any change in the methods of operating the mine that does not affect the conditions of the employee or his earning capacity . . . we would not have the right to interfere."

"Well," said one of the operators, " 'conditions of the employee' is awful wide."

"I know," replied the miners' leader. "That is the reason I used the word!"*
"Conditions of the employee" is still a very "wide" term, and the everyday work of maintaining and enforcing them keeps the mine

*District 2, *Joint Conference Proceedings*, 1903, pp. 8, 17 *et seq.*

committees and the higher union officials continually "dipping into management that is none of their business," as many operators would still put it, and gives them, in fact if not in name, an active and a significant share in the running of the mines.

Oftentimes, indeed, this comes about as the result of the simplest issues of wage rates and earning capacity. In order to ensure honest weight, for example, the men place one of their number on the tipple to check and verify the weights as taken and recorded by the company's weighman, and this checkweighman often becomes the spokesman of the men on other issues as well. Many grievances, however, require more than this informal handling. The enforcement of the scale of payment for dead work, for example, calls for representatives of the men who know the complicated set of customs and contract clauses that govern the matter, and who can go to the working places where the disputes arise and check over the foreman's measurements and interpret the agreement on the spot. It is for this and for similar purposes that a mine committee is elected in each mine,—typically three working miners who are paid by the local union for the time spent in investigating grievances; and the committee's primary duties of defending the wages of the miners frequently carry it into active contact with the operation of the mine. Sometimes, even, they put it in the position of passing judgment on the efficiency of the management. "All necessary props, rails, and other supplies," say the contracts, "shall be delivered at the working places;"* and this and other clauses are intended to make sure that the company will provide the piece workers with the proper facilities upon which their earnings depend. Often these provisions are badly enforced, and the miners have to scurry around in abandoned rooms hunting for rails and timbers that have been left behind; but at least in Illinois and in the Southwestern fields it has become the accepted principle that the men must be paid for the time that is lost on account of these and similar failures on the part of the management.† Compensation is sometimes paid not only for time that is wasted in waiting for supplies but also for time lost in waiting for the boss to give proper instructions or for the company men to lay the necessary track, switches and the like, and it has been claimed in cases where machine runners and loaders have lost time because the machines were out of order. The principle has not been pushed to the point of compensating the men for all the time spent

*District 2, *Agreement*, Rule 6.
†See *Southwestern Decisions* Vol. 1, p. 14 *et passim*.

in waiting for cars—a claim that might revolutionize the industry; but, even as it is, it means that in case after case the committees must investigate the actual operation of the mine and decide whether or not the company has "done all that could reasonably be done" to give the men work. It has been pointed out that "the genuine interest of the (British) miners in the problems of mine management (is) partly traceable to their piece-work system;"* and here in a similar way the claims of the American miners for the facilities with which to earn their wages frequently place their representatives in the position of prosecuting their employers for inefficiency in the courts of the industry.

The enforcing of the square turn also leads to similar controversies. Here again there is no such far-reaching intention. The chief purposes of the rule are the simple ones of preventing favoritism in the supply of cars on which the men's earnings depend, and of preventing men who might be favored in cars from returning special favors to the company by working for less than the scale rates. Nevertheless its day-to-day application sometimes puts the committeeman in the position of an unofficial motor boss. "They couldn't keep a square turn in one part of the mine with the other," reported one

*Goodrich, *The Frontier of Control*, p. 164.

motorman at a union convention. "They went to the mine foreman about it and he told them they couldn't do anything about it, it is up to the mine committee. So whenever one side of the mine got behind, the next day they (the committee) sent that motor over in the other side and give them a trip or two to help them out and we hain't no trouble getting a square turn."* The question of permitting exceptions to the rule, moreover, involves the committee in much more complicated questions. The union recognizes that there are certain conditions,—the need of opening up new territory to provide working places or the danger of losing coal in a "squeeze" of the roof—that justify favoring one part of the work at the expense of the others; and in these cases a "free turn" is allowed under certain safeguards. It is in judging when such exceptions are really necessary, and when they are simply cloaks for favoritism, that the committees come into most direct conflict over the actual planning and lay-out of the mines. How this happens may be illustrated by a story from a district in which the union's privileges in the matter are not very firmly established. The local union at Nanty Glo, Pennsylvania, had passed a rule that no man should load coal on the days that the mine itself was not working, with the

*District 2, *Convention Proceedings,* 1914, Vol. II, p. 488.

proviso that "the mine committee had the full rights and privileges to order the men to go to work on idle days" if there was danger of a squeeze, ("since no man wants to see anyone losing anything"), "or if the mine committee saw that it was necessary to advance this or that place." One of the companies attempted to work certain headings on idle days, the committee ordered the men not to go in, and the members of the committee were discharged for interfering with the operation of the mine. In the arbitration case that followed, the committeemen pointed out that they had not been consulted and claimed that the advancement of the headings in question was not in fact necessary for the development of the mine. When they started to present evidence on the latter point, however, the operators' commissioner cut one of them off with the remark that he "didn't need to qualify as superintendent of the mine" and asked one of the others if he presumed to "set up (his) opinion as to the necessity of certain development" against that of the superintendent. "I would," declared the committeeman. "He has the maps under his hand, but I can find my way and notice the development . . . from my own knowledge. I have a practical idea of . . . the progress and what effort has been made to develop in previous times." The ques-

tion of the committee's right to concern itself
with such matters was vigorously debated. The
foreman hotly declared that he "didn't know it
was the rule for the mine foreman to go to the
committee and ask what to do." The union's
representative argued that without such consultation the principle of the square turn could not
be enforced. The umpire to whom the case was
finally referred declared that the committeemen
were not "modest" (though he gave them back
their jobs) and affirmed the company's absolute
right, both under the contract and under what
he called "the common sense relations between
man and man," to advance its headings on idle
days if it believed it to be necessary. "Whether
such belief is right or wrong," continued the decision, "is not an official concern of the mine
committee. The management of the mine is
in no way in their hands."* But the rights that
are denied by this particular decision are recognized by the agreements in several other districts—one expressly provides that miners shall
work on idle days only "on the order of the mine
committee . . . except in urgent cases"†—and
over a very large portion of the industry the
"condition" of the square turn does in fact lead

*District 2, *Arbitration Proceedings* and *Arbitration Decisions,*
April 15, 1919.
†District 12, Subdistrict 1, *Agreement,* Clause 11.

to a recognized right and practice of consultation over questions of development.

This same feeling that lies at the basis of the square turn clause—that every man "ought to get a fair share of the work that is going"—runs also through other elements of union policy. The industry as a whole characteristically meets a depression not by drastic reduction of the working force—sometimes the total force actually increases—but by spreading what work there is over a great number of men; and in this situation also the union presses vigorous claims for its equal division.   If one section of a mine is to be shut down, the union frequently demands that the men who work in it shall share equally in the work of the other sections; if the mine closes altogether, the union is likely to claim work for the men in other mines of the same company; and there is even a clause in one contract that provides that the miners of nearby mines of *other* companies "may at their option" —not the companies'—"share work with those thrown idle, either by doubling up in the working places or some other manner mutually agreeable."* Here again the principle brings the union into contact with vital questions of management, and in working out its application the representatives of the men often come very near

*District 12, *Agreement*, Clause 12.

to exercising "the right to control and distribute the jobs."

A similar right is sometimes exercised by the union even on other grounds. In certain regions, particularly in the Southwest, the union is quite literally the distributor of jobs, under local rules and customs that provide that the operator shall hire men in the order in which they appear on the "applicant lists" of the local union. "With tonnage men the burden of proof rests with the operator to show why the man is not to be hired, and with day men the burden of proof rests with the union to show that they are capable of performing a day's work."* This practice extends over only a relatively small part of the industry, but over a much wider area the union exerts a considerable influence over transfers from one job to another and over the customary promotions from extra fireman to fireman, from trip-rider to motorman, and the like. "It was put in the hands of the bank (mine) committee," reads a familiar sort of record, "and they talked to the superintendent . . . and put the man in that belonged in."† Usually this action is based on the principle of seniority—it is the youngest man in point of service that "belongs"

*From unpublished material collected for the U. S. Coal Commission.
†District 2, *Arbitration Proceedings*, Sept. 27, 1915.

off if the force is to be reduced; but occasionally
a contract clause carries the significant provision
that disputes over competency are to be settled
jointly by the mine manager and the pit committee;* and there are even cases in which the
foreman leaves the question of the possession of
a particular job entirely to the mine committee
to settle according to its own judgment.†

Very often, also, the especial dangers of coal-
mining lead to the exercise of independent judgment by the representatives of the men.   Under
the agreement in the state of Washington, a
safety committee at each mine, composed of the
president of the local union, the mine superintendent and a third member chosen by them, is
expressly commissioned to make regular bi-
monthly examinations of the mine and "to investigate all serious and fatal accidents."‡   In
most districts, however, activity of this sort is
without official sanction, and sometimes an arbitration board takes occasion to remind the union
that "the safety of the mine rests with the mine
management and not (with) the committee."§
But whatever the contract position, story after
story from the mines leaves no doubt of the frequency and vigor of committee action on ques-

*District 12, Subdistrict 4, *Agreement*, p. 6.
†Archbald, *The Four-Hour Day in Coal*, p. 104.
‡District 10, *Agreement*, 1924, Sec. 20.
§From unpublished material collected for the Coal Commission.

tions of ventilation and safety. If a group of
men have quite work on account of bad air, it
is often a matter of course that the mine foreman and the mine committee should go together
to inspect the places in question; and such a
tour is quite likely to begin with the boss down
on his knees in the coal dust sketching the ventilating system—at the committee's request—as
the beginning of a long discussion of remedies.
If the men are afraid that a certain creek will
break in and flood the mine, if the airshaft that
serves as an emergency outlet from the mine is
slippery with ice, and in countless similar cases,
it is the mine committee that fights out the issue with the management, and often it is the
committee or the local union that sends for the
state mine inspector to settle the dispute. If
there is danger from men crowding into the
cages in the morning, it is likely to be the mine
committeemen who take upon themselves the
job of seeing to it that the men line up instead.
If the hoisting engineer raises cages of men as
recklessly as he would hoist cars of coal, it may
be the mine committee and not the management
that successfully disciplines him. And occasionally a committee takes positive and dramatic charge of a dangerous situation. One
story that comes from Herrin, Illinois, illustrates the point,—a different sort of story from

most of those that have made the town's name known. At one of the mines there had been a small gas explosion. "The following morning it was decided by the Pit Committee that men should not risk their lives in the mine unless the operator would place three good men on watch at the old workings." The company did this under protest. A week later there was a sudden alarm, the men were ordered to put out their lights, and the bosses led them to the shaft bottom by the light of their safety lamps. The manager directed that one mule should be sent up along with every cage full of men. The order "almost started a riot as the men felt men's lives were of more value than mules," and the men were finally sent up first. A few days later it was proposed that the mine should be opened again. The committee insisted upon making its own inspection before the men were allowed to go in. "They went into the old workings and were compelled to wade in water up to their necks to get to the lower part of the mine . . . (They) spent two days down the mine making this examination, and most of the time they were over their hips in the water. A proper cut was finally opened through which the air current was forced and the gas and black damp cleared out of the mine." "It cost the local union $64 for doing work that the operator should have paid for," complained some of the men, but the committeemen "felt sure that their caution"—and their independent activity—"had saved the lives of many men."* Here again, as in the issues arising from wages and the distribution of jobs, the defense of the "conditions of the employees" had led the representatives of the men into vigorous "dipping into management."

In these and in other ways, the local representatives of the union exert a considerable degree of control over the day-to-day running of the mines, and many of the issues raised in these cases are carried farther in the activities of the higher branches of the union. The question of safety has led the organization into many legislative campaigns for the adoption and the strengthening of mining codes. The principle of the fair division of work is often carried further in claims for "equal running time" as between mines of the same company and in protests against practices that prevent the equal distribution of railroad cars between different mines; and the famous demand for the thirty hour week was in large part based on the feeling that an industry that on the average provided little more than thirty hours' work a week ought to share that work equally among all the men

*From unpublished material collected for the Coal Commission.

engaged in it.   The right of consultation over
methods of work, moreover, becomes of great
importance when it is applied to the question
of the introduction of machinery or of completely new methods. For many purposes these
wider policies are of much more significance
than those of the mine committees, and the attitude of the organization toward new technique
will require separate treatment in the last chapter; but the present interest lies less in these issues than in the effect of the typical local forms
of control upon the quality of the everyday life
inside the mines.

It is clear that this power of the mine committee has a twofold effect on the freedom of
the individual miner.   In the first place, it cuts
across the indiscipline of the mines, perhaps
paradoxically, with a new group discipline. For
the enforcing of these conditions involves control by the representatives of the group not only
over the company but over the individual worker as well. Sometimes this takes the form
merely of persuasion, and the committeeman
passes on the traditions of the union in much
the same way that the other traditions of the
miners are passed on in the talk of the gob pile.
This is illustrated by another revealing story
from the Wieck Diary. The boss was trying to
get a certain miner to do a bit of company work

for nothing on the ground that he had just been treated too liberally in a dispute over dead work.

"The digger, being a little soft," so the committeeman's account runs, "took it seriously and did not answer.   The boss continued, 'Well, what do you think about it?' The digger stammered, 'I—I don't know.' 'You don't!' I exploded, and then on second thought continued, 'Well, damn it, I won't interfere,' vexed as I was at the man's denseness and ignorance of the contract. Then thinking again, the poor devil needs a little help, I said, 'Huck, listen to me.   If the boss tells you to do anything, you do it. He's the boss.   If he comes in here to tell you to throw your shovel in the gob and load coal with your hands, you do as he tells you, and if at the end of the day you think you haven't made enough money because of your instructions, and if you can't settle it with him, bring your complaint to us.'" "The man looked at me a little queerly after this lecture," ends the narrative, "and I think he got the point."*

Sometimes, however, the men do not "get the point" so easily, and the union adopts other forms of pressure.   When the eight-hour day had just been won, the new condition was con-

*Wieck, op. cit., pp. 13-14.

tinually violated by certain piece workers who insisted on coming to work long before daylight "with a lantern in their hand." "There seems to be no way," reported a union delegate, "to stop the men from going to work anywheres from twelve o'clock (midnight) to seven;" but another answered that in his local they had "caught four brothers working thirteen hours" and had tried to settle the question by fining them fifteen dollars apiece.* The enforcing of the square turn, also, has frequently meant pressure exerted upon the union's own members as well as upon the company. Sometimes this is turned upon the men who receive special favors; "any man accepting a free turn," reads an occasional resolution, shall be considered as having violated his sworn "obligations" to his fellow-members;† and occasionally more drastic action is proposed. More often the pressure is directed against the motormen and drivers who distribute the cars unfairly. Get them into the union, said one delegate in the days when the organization was weak, and there'll be no trouble about the turn. And when the transportation men are in the union, it is sometimes proposed that they should be fined if they fail to give a square turn. "A man who will give mine

*District 2, *Convention Proceedings*, 1906, pp. 23, 25, 28.
†*Ibid*, 1911, pp. 488-511.

cars for beer is not much of a man," argued a
rank-and-file delegate; why shouldn't the union
discipline him?   Higher officials of the union, to
be sure, sometimes point out that the drivers are
company men and therefore "under the bosses"
and can hardly refuse to take the cars where
they are told;* but it has been shown that in
actual practice they have great latitude in the
matter, and there is no doubt at all that the
union influence exerted directly upon them has
a powerful effect in enforcing the rule.   And occasionally this power is even used not in keeping the turn square but in enforcing some other
matter of group discipline; a man's turn may be
stopped, for example, to force him to join the
organization or to obey its orders.   "Back in
our locality," said one miner, "the day man is
the best man we have got to keep the diggers in
line."†   The effectiveness of the discipline that
the union can exert in this and other ways is
recognized in the frequent pleas of the operators
that the union should help, as it sometimes does,
in enforcing this or that rule of the company's;
and occasionally the union itself gets rid of a
man whom it will not permit the company to
discipline but whose poor work is a drag on the
other members.   In all this there is clearly com-

*Ibid, 1910, pp. 481-495.
†Ibid, 1916, p. 875.

pulsion, and a typical picture of it is that of the mine committeeman standing at the drift mouth ordering men not to work on an idle day. In his testimony in the case already discussed,* Committeeman Varish told of a dramatic encounter one morning with a group of men who wanted to go to work in spite of the union rule:

"One of them came in there special and he commenced to kick to the assistant mine foreman . . . He says, 'What is the matter, a fellow no can go in on idle day?'

I says to him, 'Listen here, didn't you insist on these rules to be drawed up?'

He says, 'Yes.'

I says, 'What in the name of God do you want to go in and tell that mine foreman you can't go in? Go ahead in, then . . .' I says. 'If you fellows wants them rules reversed,' I says, 'call a meeting . . . I am not objecting against having them rules wiped off;' and I says, 'Let every man go into the mines if they have any work.'

'Nothing doing, we don't want the old conditions we had before we had those rules.' That is the statement they made."

The encounter illustrates two aspects of the effect of workers' control on the working life. In defending the "conditions of the employee"

*See above pp. 81-83.

and in protecting its members from the rigors
of the companies' discipline, the union imposes
a new and occasionally rigorous discipline of its
own; but in the same process it gives to its members an important sense of group control.   Here
in this story the rules enforced by the committee
are clearly felt as limitations on the freedom of
the individual worker.   But they are rules of
which it can be said more literally than of those
of most democracies that the men themselves
have had them "drawed up" and that the men
themselves can "call a meeting" to "wipe them
off;" and at the end of the argument the group
of miners come to the point of admitting that the
committee is doing their business and really enforcing their will. In the great range of union activities, moreover, this second aspect—that of
control *by* the group—is much more conspicuous
than the aspect of control *over* its individual
members.   The range and vigor of this activity,
to be sure, vary greatly from district to district
and from mine to mine.   "When you haven't
got a live mine committee," said one union delegate, "you know how we suffer, you know how
the grievances creeps in."*   And just as there
are some union mines without "live" committees, so also there are a number of non-union
mines that permit mine committees of varying

*District 2, *Convention Proceedings*, 1916, p. 449.

degrees of independence.*   But on the whole vigorous committee activity is a characteristic of the union mines, and in most cases it is to the union and its officers and committees that the men turn for the carrying out of their wishes. "They go to the committees" for everything, the operators sometimes complain; certainly they do go to them on a surprising range of subjects, —from matters of a stolen pair of shoes or a broken shower in the wash-house, or the troubles of a pair of buddies who are temperamentally unsuited for sharing a place, to questions of bonding the payroll of a company in danger of bankruptcy or matters of life and death as in the story from Herrin; and it is in providing these agencies for group activity that the union exerts its most direct and most distinctive influence on the quality of the working life.   It is the union that gives to the democracies of the gob piles something of the feeling and practice of a collective freedom.

These chapters have attempted to describe the working life in the soft coal industry in terms of the technique of the mines and also of the traditions of the miners, for by the standards of other industries both mining and the

*See *Labor Relations Report*, pp. 63-65, and the voluminous literature on the "Rockefeller Plan," including Selekman and Van Kleeck, *Employes' Representation in Coal Mines*.

miner are more than "a little peculiar." It may
be that in lumbering the methods of work sometimes leave the individual as much to his own
devices as in coal mining. It may be that in the
printers' chapels and elsewhere there are groups
of workers that cling to traditions even older
than those of the gob piles. And in the clothing
trades at least there are union shop chairmen
who exercise a group control over the working
arrangements that is no less vigorous and even
more conscious than that of the mine committees. But there are not many industries that are
like soft coal mining even in one of these
respects, and there is surely no industry, unless
it be anthracite, in which all these things are
combined. The type of working life that results is unique in American industry.

On the side of technique the key to the peculiar nature of this life lies in the way in which
the workers are scattered by ones and twos in
the many miles of dark tunnels that make up
the underground workings of a mine. Almost
of necessity, therefore, the working arrangements place a quite old-fashioned reliance—for
better or worse—on "the skill, dexterity and
judgment of the individual workman." To the
independence that comes from this source, the
traditions of the miners add other elements. It
is the organization of the mine that determines

how rarely a miner shall see his boss, but how he shall act when he does see him is also a matter of the custom of the trade and the orations of the gob pile. These traditions are of the more importance because many times they are crystallized in the policies of the union, and the activity of the organization not only defends the miner's independence in regard to his own work but adds to it also something of a new collective control over the running of the mine.

Out of these two sets of influences, then, comes the freedom of the miners or the indiscipline of the mines. It *is* indiscipline, certainly, as well as freedom; if "the mine is a friendly place," as some of the reforming engineers admit, it is partly because bosses are likely to be easy-going where the piece-workers themselves stand so many of the costs of inefficiency and loose discipline; and some of the miners' privileges are hardly more than signs of the general slackness that goes with chronic underemployment. To one idealistic leader of the union, in fact, who sees in the idea of workers' control the promise of a larger freedom for the miners as a group, the mere freedom from being bossed that arises out of this looseness of mine organization seems little better than that of the sweated home-worker. The work of mining must not be idealized; it is rough and dirty and arduous, and it is carried on under conditions of semi-darkness and stale air and constant danger; but even among rough jobs there are significant differences in the quality of life and discipline; and if the occupations of modern wage-earners were to be "graded," as John A. Hobson suggests, "according to the measure of the scope of skill, judgment and self-determination" that they afford to the individual,* surely there are very many that would be placed below those of the mine workers. "The miners of Luzerne," said the delegate from a Pennsylvania local, "do feel for more liberty than money."† Their friends among the engineers and economists may well point out that what they have now is precisely "more liberty than money," more freedom on the job, that is, than opportunity for a good life away from the job; but independence is a rare thing in industry and a thing that draws many men back from the factories to the mines. "If there's anything that gives me the willies," explained one of them, "it's a boss standing and looking down my shirt collar;" and the relative freedom of the miner in this respect is enough to justify the claim that there are human values in the present type of working life underground that are worth careful consideration in com-

*Incentives in the New Industrial Order, p. 31.
†District 2, Convention Proceedings, 1920, p. 31.

parison with those of the *new discipline* that prevails in modern factories and that threatens or promises to replace the *old indiscipline* of the mines.

PART II

THE NEW DISCIPLINE

CHAPTER IV

"THE ONWARD SWEEP OF THE MACHINE PROCESS"

IT IS clear that the type of life described in the preceding pages is in sharp contrast with the new discipline* of the modern factory. While the isolated miner sets his own pace and comes and goes pretty much at his own time, the factory hand works under the boss's eye and often at the machines' speed, and his hours are rigidly bounded by the whistle and time-clock. Nor are these the only points of difference. Mining is still for the most part a hand trade, and the few types of machines that have so far found their way into general use in the mines are of the sort that the individual worker guides and controls. The mine motorman swinging a trip of cars around the sharp curves and through the dark haulageways underground is by no means to be taken as the counterpart of the typical machine worker outside. For in modern factories, "most workers are not called upon to guide machines. They follow them."† "The central fact" is no longer "the individual workman;" his share "is that of an attendant, an assistant whose

*The phrase is taken from the title of a chapter in *The Town Labourer* by J. L. and Barbara Hammond.
†Adolph Bregman, "Monotony and Industrial Unrest," *Survey Graphic*, Feb., 1923. (The writer is editor of *The Metal Industry*.)

duty it is to keep pace with the machine process."* In the ordinary case, these jobs of "following" and "keeping pace" are largely repetitive and require few of the petty decisions that make up so much of the miner's working life; and in many plants the very motions are standardized, and "planning" and "performing" are divorced, either by the minute instructions of scientific management or still more strictly by the use of machinery so automatic as to be quite literally fool-proof and judgment-proof. It is the "American automatic tool,"† in fact, that makes possible the greatest simplification of tasks; and it is along the famous conveyor lines in Ford's Highland Park plant, where the man who puts in the screw does not screw it down and where the moving belt itself "is boss" and pace-maker, that the principle of "leaving as little as possible to the whim or choice of the individual workman"‡ is carried to its farthest point with a "division of labor far greater"—so writes an admiring engineer—"than the old economists ever dreamed of" and with a regimentation of life far greater than the older critics of the factory system ever imagined.

*Thorstein Veblen, *The Instinct of Workmanship*, pp. 234, 306.
†See an article under this title by Ernest F. Lloyd, *Journal of Political Economy*, Vol. XXVII, pp. 457-465.
‡H. L. Arnold and F. L. Faurote, *Ford Methods and the Ford Shops*, p. 153.

Most factories, to be sure, are far from this extreme of the new discipline, and throughout industry there is more informality than the letter of its regulations would suggest; but when all allowances are made, the life in the mines is very unlike that of the ordinary mill; and it is so utterly different from that of a mass-production plant like Ford's that it is hard to believe that the two are continuing to exist side by side in large-scale industry. Nor in fact is it likely that they will. And there can be little question of the probable direction of the change. For surely no one is looking to the mines of to-day to see how the automobile factories of to-morrow are to be run, but many are looking to Ford's for ideas for the mines of the future; and the more unlike coal-mining an industry is in these respects the more confidently it is regarded as showing the way for other industries to follow. No one doubts that machinery will continue to spread, and it is a commonplace both in the pamphlets of the I. W. W. and in the journals of the engineers that "The Onward Sweep of the Machine Process" is breaking up craft after craft and doing away as certainly with the earlier machine trades as with the handicrafts themselves. Few will question, moreover, that for the great body of workers this means a working life with little need and little

opportunity for individual planning, and that
for many workers it means tasks so utterly de-
void of any possible interest that they are sometimes defended on the ground that the man may
perform them in a sort of "reverie" while his
mind is on something else. "Those who dis-
approve of the whole scheme of a planning department to do the thinking for the men," wrote
Frederick W. Taylor, "must take exception to
the whole trend of modern industrial development;"* and this principle is more and more
being built into the very structure of the machines themselves. The process is by no means
an even one, and there are various cross-
currents, but it is increasingly clear that the
main tide of industry is setting strongly away
from the old indiscipline and toward the new
efficiency and the new regularity and the in-
exorable regimentation of the modern factory.

This movement is perhaps the most dramatic
fact in current industrial history, and even the
coal industry has not been completely un-
touched by its influence. Mining methods to-
day, old-fashioned as they seem, have nevertheless changed considerably since the older days
of the pick miner's craft; and there are to-day
the unmistakable signs of a great hastening of
the process. The former comparison has re-

*Shop Management, p. 146.

cently been made easier by the exploration of a mine that flourished in the days before the Civil War, and its description in the *Coal Age* gives an opportunity to mark some of the more conspicuous differences.* The mine in question was a well-known producer in its time—it may even have supplied the coal for the famous race of the *Robert E. Lee*—yet what remains of its equipment looks curiously archaic to modern eyes. The nailless wooden cars, the wooden rails without trace of metal, and the great wooden bull-wheel by which cars were let down its gravity slope present a quaint appearance beside the steel of modern mine equipment and the powerful drums and motors of modern hoisting machinery. The mine mule has largely been replaced by the electric motor; simple ventilating devices like the furnace that heated the air at the bottom of the old shaft have long since given way to high-power fans; and in these and many other items of equipment coal-mining has gone far on the road to modernization.

There are other elements in the account of the old mine, moreover, that point to changes that have affected the life at the face much more closely than these. When the engineers who visited the old mine reached the shaft bottom,

*E. W. Davidson, " 'Befo' de Wah' Mine Yields Curious Relics," *Coal Age,* Feb. 21, 1924.

they were first of all struck by "the marks of skillful pick work." "The ribs* of that main entry running back from the shaft," they reported, "were as true and clean as a carefully concreted entry of to-day;" and their whole story emphasizes very strongly "the evidences of geometric exactness" and the evidences of individual skill that "marked everything they saw." Here is a picture of a real and important change. It is no longer on the eye of the miner that the operator relies for "geometrical exactness" in the mine. The pick itself has been replaced by the cutting machine in the greater part of the industry; and even where the hand work survives, much of the tradition of skill and much of the old-time miner's pride in his smoothly-sheared ribs or in his perfectly-arched headings has become a thing of the past.

The description, moreover, adds still another significant element to the change. The men who worked in the Bell Mine were all from the mines of Wales and the County Durham. "According to local history, the cross entry on one side led into workings operated . . . by Welsh miners and the other into Durham territory." The two groups "were reputed to have scoffed at each other's skill as miners and to have maintained such keen rivalry that it periodically

*The side walls.

broke out in free-for-all fights both in the mine and outside. Standing at the junction of the main heading with the side entries, it was evident to the three explorers of the mine that the two entries had been driven by different types of miners. The Durham entry had been sheared (cut vertically) on the right hand and shot from the left while the Welshmen—whether they were all left-handed or not—had sheared on the left and shot from the right." British miners such as these, whose work shows unmistakably when examined seventy years afterward that they had learned their craft in the mines of the old country, were a dominant influence throughout the greater part of the industry in the earlier periods. The mines of to-day on the other hand are peopled much more largely by men of the new immigration and their sons—Italians, Poles, Hungarians and many others—whose old-country tradition is one not of the mines but of the farms. More than eighty per cent of the British-born miners who were studied by the Immigration Commission in 1910 had worked in the mines before coming to America; but the great majority of the mine workers of the new immigration had been farm laborers or peasant farmers in the old country, and only some ten per cent had ever worked in mines.* The same

*Immigration Commission, *Immigrants in Industries*, Vol. I, p. 43.

contrast, moreover, would doubtless hold at any period at which the two streams of immigration could be compared. It is in its personnel that the industry has perhaps changed most completely since the days of the Bell mine,—a change from the old to the new immigration, and from men who came knowing their trade to men who come to our mines as greenhorns.

Along with these changes, moreover, and in part because of them, there has come something of an increase in the amount of supervision. In the old days, the foreman might set the sights for a man's place when it was started and then never visit it again, except perhaps to make sure that he was not weakening the roof by driving the room too wide. A traditional story tells what happened when a boss tried this method of work with a certain foreigner of the new immigration. He had given the man his place and had hung from the room the two sights that were to give the line of direction for the room. Next day another miner happened to come by and saw that there was only one sight hanging up:

"What's the matter, John? Where's your other sight?"

"Me keep it safe," explained the greenhorn. "Use it when that one gets broke."

Men like this one clearly needed much more than

the customary direction, and were very likely willing to take more than the customary bossing; and no doubt it is their coming to the mines that has been largely responsible for the more regular visiting of the working places. In the meantime, also, more systematic methods of laying out the mines have somewhat concentrated the work and made it easier for the foreman to make his rounds; the opening of deeper and more gaseous mines has in special cases called attention to the need of frequent inspection; and the movement throughout the industry toward stricter standards of safety has tended in the same direction. Workmen's compensation laws have done something toward "making props cheaper than men," as one miner put it, and supervision cheaper than accidents; and it is the state mining codes, passed as safety measures, that in the greater part of the industry set the current standards of the frequency of the foreman's visits. "Consider the customary supervision," comments Mr. Archbald, when this much has to be "ordered by law;"* but it is at least true that, for these and other reasons, the standard of a daily visit to the working place— though not strictly enforced—has in most coalfields taken the place of the almost unbelievable infrequency of supervision of the older period.

*Four-Hour Day in Coal, p. 36.

All these are real changes,—the revolution in equipment, the introduction of the mining machine, the coming of the new immigrant, and this increase in the frequency of supervision— and the picture of the industry would be falsified if they were left out; yet surely the striking thing is not that the coal industry has changed in these respects but that it has changed so little. The new equipment has altered the look of the mines much more than it has really affected the general quality of the work, and it has hardly touched the work of the man at the face. The coming of the mining machine, to be sure, together with the changes in personnel and supervision, marks the passing of the "old-time miner" and much of his tradition of pride and skill. Yet even these things have made surprisingly little change in the old indiscipline. An increase in supervision that has not even brought the foreman's visit securely up to a minimum of once a day can scarcely be taken as a very great step toward factory regimentation, and the coming of the mining machine has hardly altered the independence of the man at the face. The machine loader preserves his property right in his place as jealously as any pick miner of the older days, and the machine adds to his life a further element of irregularity—that of "waiting for cuts"—rather than an element of standardization.   The arrival of the newer stream of immigration has of course changed the whole color of life in the mining communities in a hundred ways.   Hungarian Grape Harvest Balls, for example, have taken the place of Welsh singing festivals.   Old-world feuds of Pole against Russian, of Slovak against Magyar, and many others, have been brought over to the mining camps.   Many a community, also, has been cut in two by hostility between native and "foreigner"—a hostility illustrated by the Herrin outbreaks in which Baptist and Methodist and Campbellite Klansmen have been aligned against Catholic and Orthodox immigrants. And the change has time after time given to the union the problem of teaching its tradition—across the barriers of these prejudices and with all the obstacles of a dozen languages—to new groups without industrial backgrounds and with lower standards of living.   Yet for the specific problems of this book, the striking thing is not the far-reaching effect of the change but rather the surprising continuity of the miner's tradition in spite of it.   The new immigrants as a whole have not shared the older miners' objection to the cutting machine.   They are perhaps a little more likely to "doff their caps in the old-country manner" when they meet "Meestaire Supaire" on the road to the mine.* But for the most part they seem to have taken over quite thoroughly from the older miners the attitudes that were customary in the rooms and on the gob-piles; and the union that was so largely founded by British miners has passed on its traditions and policies with very little change in spite of its changing membership.†

The isolated mine workings, out at the ends of their branch railroads, cut off from each other and from contacts with other types of work, have felt little of the stir of change that the efficiency movements have brought to the factories. The task of standardizing work that must be done underground and in continual advance from place to place would have been difficult under any conditions; and the pressure of part-time operation, together with the tonnage system of payment, has made it easier to throw part of the cost of easy-going methods onto the miners themselves than to spend money in the extravagant overhead of superintendence and engineering and experimentation. With the conservatism of the miners and the still more

---

*Archbald, "Mister Super," p. 1021.
†Note in Andrew Roy's *History of the Coal Miners* that the early leaders were almost all British, that a number had been members of British miners' unions, and that several had been under the influence of the great Scottish leader, "Sandy" Macdonald. On the passing on of the tradition, see W. M. Leiserson, *Adjusting Immigrant and Industry*, pp. 195-6.

significant conservatism of bosses bred in the routine of the mines and for the most part knowing nothing of other methods of work, the remark of a shrewd foreigner that "costume predominates in coal mining more than any other industry"* has long been accurate in everything but its English. Coal mining has lagged far behind other industries in the development of the new discipline; and, in spite of all the changes that have come about in the transition from the period of native or British pick miners to the day of Slavish and Italian loaders after the machine, the main outlines of the miner's freedom have remained—so far—remarkably unchanged. Yet in the last few years there have been definite signs that the cake of "costume" is being broken and that the slow change of the past is giving way to a much more rapid movement in the direction of the new discipline.

One indication of this lies in the fact that so much of the recent comment on the industry, both from within and from outside, has centered on the issue of its backwardness as judged by current standards of efficiency. Even before the United States Coal Commission was appointed, plans were under way in the Department of Commerce for an investigation that would, in the significant phrase, "apply to coal

*In a letter to *Coal Age*, March 2, 1922.

mining the tests of good management that have been worked out for shop and factory work;" and the study of "Underground Management" made for the Commission under the direction of an industrial engineer, Mr. Sanford Thompson, represented the carrying out of this intention. The report makes comparison after comparison between the "hit-or-miss supervision" of the mines and "the advanced methods of planning and control and standards of production that are being worked out to such a high degree in many industrial establishments" and argues vigorously for the "necessity of the same trend in underground management that has proved so effective in industry above ground."* Similar comments have also been made, from a vantage point somewhat closer to the control of the industry, by engineers and officials of the great coal-consuming companies that now own a substantial portion of the soft coal industry and whose own ruling tradition is that of more regularized methods of work. Mr. Arthur J. Mason, for example, an ore engineer employed by the United States Steel Corporation, has drawn a sharp contrast between "the art of ore handling" which he and his colleagues had done so much to revolutionize and the backward "art of coal mining," and has asked whether "it is just

*Thompson Report, pp. 2, 6, 15.

to the rest of us that this particular calling
should remain so far behind."*   The same question, moreover, is being raised quite as insistent-
ly by men within the coal industry itself.   Mr.
Hugh Archbald, for example, a mine superintendent and mining engineer, has stated the contrast in disciplines as sharply as any outsider,
and with greater knowledge; and his book on
"The Four-Hour Day in Coal" is a vigorous
plea for a more intelligent "engineering of work"
underground.   "In factory work such engineering is in full swing of development.   In coal
mining it is not."†   The same doctrine is also
being preached—and very ably—in the journals of the industry, and the marked increase in
news and advertising devoted to new methods
and new devices is an important indication of
the stir of change. *Coal Age,* to be sure, admits
that it is doing more than "merely mirror that
progress."   It intends to "uncover it in its beginnings and fan the faint spark of each local
development into a flame for the advantage of
industry;"‡ but there are many independent
signs that the "flame" of which it speaks is
actually spreading rapidly through the industry.

One such indication is to be found in the increasing pressure exerted upon the union on
issues of discipline and technical change.   Dis-

*Op cit., pp. 106-8.
†P. 143.
‡April 10, 1924 (Editorial Insert on Contents page.)

cussions of time clocks and new methods and new devices, and of the strict enforcement of the full-day and the lay-off clauses, have in the last few years played a more important part than before in public controversy and at the scale conferences and also in the day-to-day dealings between operators and union officials back in the coalfields. Another and a quite distinct indication of the same general movement comes in the increasing influence and the increasing confidence and professional feeling of the engineers within the industry. Coal mining is still greatly undermanned with technicians, at least as compared with metal mining, and it is still the tradition for the "practical" men to look down upon the technical men, but here too real changes are under way. The number of engineers has increased in recent years, and also the deference paid to their judgment, and the meetings of their technical associations are beginning to be important forums for the discussion and therefore the promotion of technical change. The sessions devoted to new methods of coal mining at the 1924 meeting of the American Mining Congress were a significant illustration of this; and the eagerness with which hundreds of operating men followed the discussions there—and their interest in the new types of machinery exhibited—were genuine indications of an increasing receptiveness toward ideas of change. Gatherings like that of course fail to represent the inertia of the greater part of the industry; yet there is similar talk of innovations among operating men in every coalfield—talk of Ford methods and Taylor methods and metal mining methods as well as of specific suggestions for coal mining—and talk so widespread and insistent that it is perhaps the best sign that the industry is embarking upon a period of active experimentation.

Throughout this comment, moreover, and through much of the current literature of the industry, there runs a note of expectation and of confidence in the beginnings that have recently been made. "It is generally conceded that the next three years will show more real progress in the art of mining than the last fifteen," or so the trade journals claim; sometimes, even, they prophesy revolutionary changes "a few months hence." "The industry is parting with the past," writes one enthusiastic editor, "and greatly venturing forward into the future." It is time, then, to turn to the more conspicuous of the new ventures on which such talk is based. "Now," declaims *Coal Age*, "Now for the Big Push! . . . Just now, more than ever, methods are changing."* Now more than ever, then, it is important to examine the bearing of these changes on the quality of the working life.

*Editorial Inserts on the Contents pages of the issues of Jan. 10, April 10, April 24, and May 1, 1924.

CHAPTER V

THE NEW TECHNIQUE

THE current revolution in mining methods
—"the Big Push" that the trade journals
are heralding—is hardly to be thought of as a
concerted movement. Many operators have
been seeking lower costs by many different
methods, and many engineers and manufacturers and officials have been laboring to serve
them and to make mining more "progressive"—
for the most part in considerable isolation. The
resulting innovations are naturally of many
sorts. Yet through them all there runs a clear
tendency toward bringing into the mines the
regularity and discipline of the working life in
the shops. "The best mines"—so enthusiasts
declare—"are coming more and more to be organized like factories."

Certain of these new methods involve little
change in the equipment and basic technique of
the industry. The simplest way of going about
it to make a mine "just like a shop" is merely to
bring over bodily some of the characteristic rules
and regulations of factory life. This is what
has been done in the conspicuous case of a large
coal-consuming company, well known for the

degree of regimentation in its manufacturing plants, that has recently purchased a group of coal mines. Its first move was to introduce two new rules,—the one, that there should be no smoking during working hours—an unheard of restriction for non-gaseous mines—and the other, that a man who wanted to lay off for a day or more should arrange with his foreman for a written leave of absence to be made out in triplicate beforehand. In the latter case not only the rule but also the very printed forms themselves—with the name of the parent company at the top and the headings, "Verbal Leaves of Absence Don't Go," "White Copy to the Employee, Yellow Copy to Time Department, Pink Copy stays in this Book," and so on—were brought directly from the factory. To outsiders these requirements may not seem very surprising, but the stir that they made inside the mines and in the whole region round is an indication of the revolutionary change in the psychology of mine management that they typified. The superintendent, to be sure, tells only of a "flurry" among the miners when the rules were put into effect and of a few discharges. "At first they thought it was none of our damn business," he said. "We soon showed them whose damn business it was." But months afterwards operators for a hundred miles around, in a solid-

ly non-union field, were still telling with amaze-
ment that, "——— had gotten away with a
rule against smoking;" and the comments of a
Logan County mine foreman—a foreman, that
is, in the most famous non-union county in the
world—mark even more sharply the abrupt-
ness of the change. He hardly believed the
story of the no smoking rule—"it can't mean
for the tonnage man, too," he said; and when
he heard of the absence slips—we were then
going the rounds of his mine—he had to stop
walking to express his indignation. "You don't
believe in that, do you?" he demanded. "Why,
I wouldn't ask men to do that. I wouldn't work
under a system like that."

The "smokeless" rule and the pink absence
slips, to be sure, are more symbolic than important in themselves, and there are few companies that are attempting to graft the exter-
nals of factory regulation quite so directly upon
the traditional organization of the mines; but
the tendency toward the tightening of discipline
is a very general one, and there are a number of
indications, also, of an attempt to bring over to
the coal mines—long after the original sweep
of the movement through other industries—
some of the characteristic principles of scientific
management. The authors of the Thompson
Report, for example, argue that the work should

be placed "on an incentive basis" and also that
the "efforts and knowledge (of the foreman)
must be supplemented by the work of special-
ists" both in the study and planning of methods
and in the superintendence of particular aspects
of the work.*   The actual experimentation in
these directions, however, has been very slight.
The differential piece rate and similar systems
took little root in the mines, and to-day the rival
factory doctrine of completely standardized day
work according to the methods of Ford is at
least equally influential.† Engineering analyses
of the methods of work, as Mr. Archbald points
out, are almost unheard of. Ordinary time
studies are very rare indeed, and the Coal Commission's analyses of the problems involved in
haulage control‡ were apparently a complete
innovation in the literature of coal mining.
There are, however, at least the beginnings of
an approach to the scientific management ideal
of functional foremanship.   A few of the larger
companies are experimenting with specialized
personnel and employment departments.   In a
rather larger number of cases, company safety
inspectors visit the various mines to check up on
the safety work of the foremen and their assis-

*Pp. 2, 10, 44-5, 15.
†Goodrich, "The Obsolescence of Piece Work," *University Jour-
nal of Business,* June, 1924.
‡*Thompson Report,* pp. 15-31.

tants, and sometimes their reports determine the bonuses that are given to the other officials. Sometimes also one finds an old workman going about the mines as a "shooting boss" to instruct the miners in the proper preparation of their shots—at least when there is little market for small coal—and in a few cases there is a more dignified "campaign of education" in this regard under a "superintendent of explosives" with some degree of special training. A series of experiments recently conducted by the Bureau of Mines, moreover, suggest the possibility of a much more complete substitution of science for craft skill in the methods of blasting; and the activities of certain groups of operators and of the large powder companies suggest the probability of a definite movement toward taking the discretion out of the hands of the individual miners and placing it instead in the hands of experts under the direct control of the company.* But these things are only beginnings and possibilities, and in the present state of the industry they are of much less importance as steps toward the new discipline than certain systems of management already in operation that are based on the simpler principle of an increase in the *amount* of supervision.

*See paper by J. E. Crawshaw, U. S. Bureau of Mines, and discussion at May, 1924, session of the American Mining Congress.

One of the most interesting of these systems
is that worked out by Colonel Edward O'Toole
and his associates for the mines of the United
States Coal and Coke Company, a Steel Corporation subsidiary, in and around Gary, West
Virginia.   Its cardinal feature is simply a great
increase in the number of bosses.   In the ordinary mine, each foreman or assistant has in
charge some hundred men underground.   Inside
the Gary mines there is one assistant or "cut
boss" to every twenty-five men.   This change
brings with it entirely new possibilities in supervision.   It means that it is as easy for the Gary
cut boss to visit each man's working place every
two hours as it is for the ordinary bosses elsewhere to complete their rounds every day or so,
and the requirement that the bosses must touch
off the men's shots with their electric batteries
does in fact ensure this frequency of supervision.
The change makes possible a much more accurate control of haulage.   "There's no such
thing in these mines as the driver being the boss
and taking the cars where he pleases."   A considerable part of the Gary bosses' pay, moreover, depends on the success or failure of their
sections in avoiding accidents, and the system
places upon the bosses a much stricter responsibility for safety than in the ordinary mine.
Above all, by the deliberate policy of the com-

pany, the new system means a relationship
between the boss and the individual workman
that is quite foreign to the older traditions of
the industry. One of the superintendents, for example, happened to find a newly-appointed cut
boss and an Italian loader together in the man's
room, and took the occasion for a lecture on
their respective duties. "You're boss," he told
the new assistant. "If they don't do what you
tell them, send them along out." Then to the
loader:— "Joe's boss. Versteh'? Joe's king
in here. You know, same as in Italy." There
was of course some humor in this, and the cut
bosses themselves, mainly men of the new immigration, are rather less kinglike than this suggests and do leave rather more to the individu-
ality of their fellow-foreigners than the letter
of the system calls for, but the emphasis on the
foreman's decisions is nevertheless quite unusual.
In many mines question after question is answered, "It's altogether on the man." In these
mines, the slogan is rather:—*"It's altogether on
the boss;"* and the spirit of the company is accurately expressed by one of its daily safety bul-
letins which declared that the mine foremen and
assistants "are supposed to do the thinking for
the men."

The system is unusual, also, in the degree to
which the company "does the thinking" for the

bosses as well. A good story is told of the beginning of this policy. At the time when the
first great increase in the number of bosses was
made, a certain colored loader declared:—"I'se
been with this company two weeks, and I'se
learned seven different ways of loading coal."
"That showed us," said the superintendent,
"that we were not in accord;" and the result has
come to be a highly standardized system.
Printed instructions to the foremen—instructions "superior to anybody"—are intended to
prescribe almost every detail of the work. Exact
specifications cover such matters as the method
of wiring for the electric lights that are installed
in every working place, the laying of room
switches, and even the setting of timbers in the
individual rooms. The Gary mines, in fact, are
known through much of the state as "the place
where they set posts by blueprint." The point,
however, at which the exact control by the company over both men and bosses is felt most con-
tinuously throughout the day's work is in the
device of the "estimate." Its influence begins
at starting time. Each loader knows that he
must "clean up" during the day all the coal that
has been cut in his place so that by night the
place will be ready for the next cut. When the
boss makes his first rounds, he counts the number of men in the places, figures with each man

how many cars he will get from the coal that has been cut, and adds that total as the estimate from his section. The estimates from all the sections in all the mines are telephoned to the head office, and from then on their pressure is exerted throughout the day on every boss, on every man in the haulage system—for every driver and every motorman knows how many cars he is supposed to haul during the day—and on every man at the face. At the end of the day, the results are compared with the estimate, on a daily report that carries a warning space for "Reason why estimate was not reached," and in practice the totals do come surprisingly close to the estimates made. For one day taken at random, for example, only two out of twelve mines fell below their mark, and seven mines dumped precisely the estimated number of cars.

Largely as a result of these things, the records of output per man are very much higher than the average performances of the industry, and the entire atmosphere of hustle and regularity within the mines is very different from that of more informal and less rigorous companies. The minimum requirement that is exacted of every loader is far above the average production throughout the industry; "when estimated cars do not average five cars per man for three ton cars," reads the daily report, "a reason must be given below;" and the actual records run well above that figure. Throughout the whole organization, also, there is a very sharp feeling of its separateness from the rest of the industry. The company makes it a policy to train its own bosses, very rarely taking them from other companies—"we don't want them to know anything but what we tell them"—and the same attitude is carried all down the line. Many of the men and some of the bosses have never worked in other mines, and in any case the policy is that "every new man is a greenhorn." "Sometimes a man will come and say, 'I've got my miners' papers from such and such a state, and I don't need to be told how to mine coal,'" explained one of the officials, "but we say, 'I don't care what papers you've got. You've got to go by our system here!'"

Here, then, and under a number of other companies, the whole psychology of mining has been radically altered by changes that do not involve any splitting up of the miner's job or any important innovations in mine lay-out or in machinery. The loader at Gary still performs by hand all or almost all of the varied operations of the man in the ordinary mine. He still sets his own posts, for example, if it is "by blueprint;" it is only the "thinking" that the com-

pany is supposed to do for him. But in the
meanwhile other operators have been making
changes in these respects also. A further division of the miner's work is by no means unusual.
Room track, for example, is often laid by company men; and over nearly half the industry,
as a matter either of state law or of company
policy, the touching off of the shots, though
usually not their preparation, is done by special
groups of shot-firers. The cases in which the
division of the work is put definitely on the basis
of skill are less common but more significant.
In a number of Southern mines the miner himself, as in anthracite, takes with him a "Chalk-
eye" or helper to do the less skilled part of the
work. A more systematic step in the same direction was taken by a number of operators in
the nineties who divided the work between
groups of "shooters" who were to do the timbering and the shooting and all the skilled work and
"loaders" who were merely to shovel the coal;
and, although this practice was checked by the
increase of the union's strength, union operators
have sometimes looked back to it with regret
and non-union companies have sometimes continued the arrangement. In recent years, moreover, the invention of electric or compressed air
drills to take the place of the miner's breast
augers or grip-bar drills has been a powerful

stimulus toward a movement in the same direction. In several fields the work of drilling the holes is now almost universally done by special crews who go from room to room with the drilling machine; and sometimes company drilling is supplemented by company preparation of the shots and by company timbering even in the rooms. In these cases, the skill and interest are concentrated in the work of these special crews —"There's a genuine affection," claims the advertisement of an electric drill, "in the way (the men) handle this little machine," and it is at least true that they "handle" rather than "follow" it—and the loader is reduced almost completely to the position of an unskilled laborer. But the scattering of the working places still stands as an obstacle to the spread of this practice, and the more important movement toward the breaking up of the miners' craft is coming by way of other new machines that take over much more of the loader's work and that often bring with them changes in the lay-out of the mines as well.

One of the most interesting of these devices is the *face conveyor*. Its central idea is a simple one, that of applying the common factory device of the moving belt or conveyor, already used in the outside work of a number of mines, to the task of hauling the coal from the face itself. The loader shovels onto a knee-high moving belt instead of up into a mine car; conveyors and not cars carry the coal to the heading. But to make such an arrangement possible involves drastic changes in the organization of the mines; and under the "v-wall" system developed at Norton by the West Virginia Coal and Coke Company, the introduction of face conveyors has cut the loader's job in two, changed the whole method of laying out the mine, and brought about also a less noticed but an equally revolutionary change in the whole manner of supervision and discipline. It would not be practicable to run a conveyor all the way into each of the narrow rooms of the ordinary mine for the small quantity of coal that could be loaded from it; the first necessity of the new system is to have longer faces on which to work. These are provided by an ingenious plan under which the coal is mined from pairs of faces each 75 to 100 feet long, set at forty-five degree angles to each other in a saw-tooth line. The teeth of the saw are solid blocks of coal; they point toward an area from which all the coal has been mined out; but in each of the v-shaped spaces between them the roof is held up by the support of the teeth themselves and by thickly set timbers; and the coal is loaded onto the conveyors along the two inner sides of the V. Each

of these face conveyors empties at the point of
the V into a cross conveyor that takes the coal
through a narrow passageway to a lateral conveyor that carries the coal from the whole set
of faces to dump into mine cars out on the main
heading. In this way many hundred feet of
working face are reached by a compact transportation system capable of continuous operation; the next necessity is to keep the coal from
that face pouring constantly onto the conveyors
throughout the day. This is accomplished by
giving the loader practically nothing to do but
load. He must still pick down part of the coal,
to be sure, and he must throw the impurities
over the conveyor into the gob; but otherwise
his work is simply to stand and shovel coal all
day. The many other things that the ordinary
loader does while he is waiting for cars, and with
them all the more skilled parts of the work, are
done by other workers on the night shift. A
crew of four men on each pair of faces, for example, does the cutting, moves the conveyors
forward to the new faces and timbers the space
behind them; a crew of two men drill and shoot
the holes for a number of faces; and there are
several men who shorten the cross conveyors
each night—as the advance of the face brings it
nearer to the lateral—and set up the spare sections in the next block of coal to be mined. Once

every two weeks or so, moreover, there is the exciting job of going back into the mined-out region and notching enough of the timbers so that the overhanging roof will fall and relieve the strain over the working area.

This division of the job marks a sharp innovation in the discipline of the mines. The work of the night gang, to be sure, has all of the customary variety of mine labor and calls for somewhat more than the average skill; except for the fact that even the cutting is paid by the day and that all the work is done under unusually close supervision, it involves comparatively little of the new discipline. But for the much larger group of loaders, who make up the bulk of the day force, the new system means a type of life quite unheard of in the mines. Its uniqueness lies not merely in the almost complete reduction of the job to one simple operation but in the whole set of arrangements under which the work is carried on. It is a gang job instead of an isolated one; four or five men work side by side in each of the narrow alleys between conveyor and face—digging themselves out of the great piles of coal that fill the space at the beginning of the day and cleaning up the last of the loose coal from the floor at quitting time. It is a day work job—"loading coal by the acre," as the men's phrase goes, "instead of by the ton"—

and each man has a certain length of face to clean up as his daily task. And it is a job done under full supervision,—with frequent visits from the foreman and with a strawboss in constant charge of the eight or ten men in the lighted space between each pair of faces.

This change to a more complete control over the men and an increase in "harmony" which the company attributes to "the psychology of grouped work" are in fact claimed as among the chief advantages of the system. "Note how several men work together," says an editor of *Coal Age,* "making the labor of loading more congenial than when in the room alone or in pairs." "A loader has a 'cinch,'" he remarks, "for he does not have to exert himself by throwing coal for long distances or to any considerable height."* A group of five men who were doing the job, however, were much less sure on this point. I asked them whether they liked it better than working in a room. Three of them had never worked in the mines before, and one of these thought that he would like ordinary mining better. "This," he said, "is damned hard work." Of the two who had tried both, one answered, "No, but I like the pay better;" and

*A. F. Brosky, "With Hand Shoveling 10 Tons Per Man and 26 Tons Per Loader Already Attained," *Coal Age,* Feb. 7, 1924. This article contains a complete description of the system with diagrams and illustrations.

the other explained that "you can't get an old-time miner to work on this. I wouldn't, only the other work is slack just now. I don't like shoveling all day." Union officials, moreover, have described a similar job as "a slave sort of work;" and in this case the union advocated a different method of payment—a tonnage rate to be divided equally among the group—partly with the intention of getting rid of the straw-boss and preserving for the shovelers something at least of the traditional independence of mine labor. For whether the new system is "congenial" or not, it is clear that the type of work—that of "mere shoveling"—and the type of supervision—that of a strawboss over a gang of task workers—are very far indeed from the characteristic conditions of the miner's freedom.

This conveyor system is already producing a standard output of twenty-six tons per loader per day—more than three times Archbald's estimate for the work of a loader in the ordinary mine—and is being rapidly extended, and it marks as sudden and dramatic a change in the manner of life underground as any other single innovation; yet even this does not represent the most thorough-going application of machinery to the miners' work. The company that has developed the face conveyors hopes finally to put machines instead of men to the work of shoveling onto the conveyors, and it is around "the burning question of machine loading"* that the most widespread experimentation is centering. "Around the corner of almost every gob-pile in the American coal fields," declares an enthusiastic editor, "somebody has been quietly developing a (mechanical) loader;"† and a great variety of machines have resulted,—from a simple bottomless scoop that drags the loose coal along the floor to a great battering ram of an entry driver that charges boldly into the solid face of the seam. Many of the attempts have been failures, to be sure, and some of the successes have been carefully kept under cover; but mechanical loading is fast emerging from the purely experimental stage. "This change in the method of mining from men to machinery, while yet in its infancy, is (already) an actual accomplishment," as the Coal Commission recognizes,‡ and it is by no means too early to describe the leading types of machines and to give some indication of the important changes that they are bringing in the life and discipline of the mines.

---

*A. F. Brosky, "Interpretation of Progress in Mechanical Loading," *Coal Age,* Jan. 17, 1924. This article gives a full description of the various makes of machines.

†*Coal Age,* Feb. 28, 1924.

‡*Thompson Report,* p. 6. A large portion of this document is given over to detailed studies of the performance of mechanical loaders and to the advocacy of their introduction.

The simplest of these new devices are the *scraper loaders,* of which there are between fifty and seventy-five at work in the bituminous mines.* Their essential feature is a v-shaped scoop, without top or bottom, attached by a system of ropes or chains to an electric hoist or winding engine. One man guides the scoop into the loose pile of coal; another operates the hoist; and by its power the scoop is dragged along the bottom to dump its load either onto the end of a short conveyor that loads the car in the room itself—as in the Jeffrey type of scraper —or else up over a chute into a car out in the entry—as with the Goodman machine. In the first case, the machine may be operated with these two men only, though frequently a third is added to shift the cars to and from the conveyor. On the other hand, the machines that load on the entry do away with the need for mine car haulage in the rooms—a great advantage in low coal, but they require a slightly larger gang than the others, with an extra man at the face and one or two more at the chute. Under some companies these men at the face do the timbering and a number of the other jobs of the ordinary loader—a Jeffrey advertisement even makes a point of how many things they

*These and the following figures are rough approximations based on the estimates of manufacturers and other authorities in the summer of 1924.

may do while waiting for cars—but in other cases there is a much more thorough-going division of labor. The more striking changes in the discipline of the mines, however, as well as the greater number of successful experiments in mechanical loading, have come not with these devices but with the introduction of more complicated machines that do away with the detached scoop and concentrate the work of loading in a single compact mechanism.

Of these larger machines, there are a few that carry over from the scrapers the principle of the scoop but mount it instead on a machine that is very much like a smaller form of the ordinary steam shovel, except that the power is either electric or compressed air. There are a number of these *shovels* in operation in the coal mines, particularly in the Western fields; but for the most part they require too much headroom for the ordinary seam of coal; and the most widespread development in mechanical loading has come with still another kind of machine that operates on quite a different principle. Of this third type, which the Coal Commission describes as the *"double conveyor,"* there are already more than two hundred machines in operation; and one manufacturing company is said to have twenty-five orders a month for months ahead.* Its essential feature is an arrangement of two or more moving belts or conveyors,—one to lift the coal from near the level of the floor as the head of the machine is moved up to the pile of coal, and another to carry the coal back over the machine and discharge it into a mine car behind. So much is common to eight or more makes; the most striking differences among them are in the devices used to get the coal started on the lower or elevating conveyor. The Myers-Whaley throws it straight backward with an automatic shovel that breaks into two sections so that its swing does not carry it above the height of the machine; the machine in most common use, the Joy, "fights the coal onto the conveyor" with a pair of "clawlike" revolving arms; and there are as many other devices as there are other makes of machines. For the present purpose, however, these differences are of less importance than the differences in the number of men that serve the machines and in the nature of their work.

The actual running of the machine,—the manipulation of the controls that govern the forward and sidewise motion, the swing of its loading and discharging booms, and the operation of its conveyors—is in all cases a one-man

*The Joy Machine Company authorizes the statement that up to June 1, 1925, it had sold approximately 295 double conveyor machines.

job; and the makers of the smaller machines claim that this single operator is the only man needed in attendance on the machine, except for the work of supplying it with cars to load. "Its introduction," says a Joy advertisement, "is virtually the same as supplying a single miner with a machine to supplant the picks and shovels of from fifteen to twenty miners." In actual operation, however, the picks and shovels do not always disappear so completely; and even with the smaller machines there is very often a man to pick down the coal in addition to the helper who shifts the cars behind the machine. The most successful of the larger machines, the Coloder, calls for several other helpers, with from two to four men at the head of the machine to break down the coal and to set the jacks by means of which the loading boom is swung from side to side, and with a boss who spends the whole day with the gang in the dust of the machine. With all of the double conveyors, also, the timbering, tracklaying and cutting and also the blasting of the coal must be done before the machine enters the place; and the last requirement adds a drilling-and-shooting crew of two men, or in some cases separate crews for each operation, to the many scattered groups that already perform the company work in the mines.

There is one group of mechanical loaders in

use, however, that does away with these specialized drillers and shooters, and with the cutting
machine men as well, by combining the cutting
and the loading in a single operation and by
tearing the coal out of the solid seam without
the use of powder. The best known of these
combined *cutting and loading machines* is the
Jeffrey Entry Driver, which undercuts the coal
and shears it at both sides with three great sets
of cutting chains and at the same time punches
down the coal with a battering ram set with pick
points. The broken coal falls onto a conveyor
at the bottom of the machine and is carried back
to the mine car by a double conveyor system like
that of the simpler machines. By this ingenious
combination of processes, the cutter and loader
is enabled to press into the coal at a rate much
greater than that of the smaller machines and
quite beyond comparison with the foot-by-foot
advance of hand-driven places. One of the
Jeffreys, in fact, with only a three man crew, is
recorded as driving forward fifty-three feet in
seven hours. The machines of this type, however, represent a very considerable investment,
—some twenty-five thousand dollars apiece as
against between five and ten thousand for the
commoner makes of double conveyors; and so
far their chief use has been in the sort of work in
which speed is more important than cost. Almost

all of the fifteen or more machines that are now in operation are driving only the headings or entries, the narrower passages that open up the coal as a preliminary to taking the bulk of the output from rooms and pillars. But there are several operators who are carrying the process further and adapting the cutters-and-loaders to the work of the quantity production of coal—one such machine has already loaded as much as six hundred and twenty tons in twenty-four hours and fifty-five tons in a thirty minute spurt*—and there are a number of mining men who see in these combined machines the possibility of the compact mechanization of almost all the work inside the mines.

Of all the machines, these big cutters-and-loaders make the sharpest break from the customary organization of the industry, for they carry the process of mechanization to the farthest point and they do away most completely with the characteristic isolated labor of the mines; but with any form of mechanical loading, the revolution in the character of the life is a very real one, and in many cases its effects extend considerably beyond the immediate change—striking as it is—from hand shoveling to the machine jobs at the face. For although the scraper loaders are expressly advertised as not

*U. S. Coal and Coke Co. Reported at 1924 Mining Congress.

requiring "any sweeping change in the methods of mining," and although even some of the larger machines have been operated with fair success with very slight changes in the rest of the organization, nevertheless many operating men are beginning to discover that they cannot expect the inventors and manufacturers to develop machines that will "step into the coal mine and do its work without making any change whatever in the methods and organization hitherto in use."* The argument is a very simple one. The newer type of Joy loader, for example, is sold under the positive guarantee of a capacity of two tons a minute, and in tests this rate is frequently exceeded; yet in actual operation an output of two hundred tons in an eight hour shift—instead of the theoretical nine hundred and sixty—would be considered an excellent day's work. The reason for such discrepancies, moreover, does not usually lie in any breakdown of the mechanism itself; it is simply that the machine—like the miners it is replacing—must be idle a good part of the day. One series of time studies made of a group of Joys in fairly successful operation shows that they were actually loading only three-eighths of the time.† The Coloder apparently holds the records for

*Howard N. Eavenson. In discussion at 1924 Mining Congress.
†David Ingle, "Making Coal-Loading Machines Work Successfully in Indiana," *Coal Age,* Jan. 31, 1924.

both the largest and the most consistently maintained tonnages, and its operation is so standardized that the Pocahontas Fuel Company
sends each machine crew inside with a hundred
cars as its day-in-and-day-out stint; yet even
with these, a series of performance records that
show an average output of over three hundred
and fifty tons in eight hours show also that the
average loading time was only fifty-six per cent
of the day.* The explanation is that under
ordinary mining conditions every one of the
simpler machines can load more coal than can
be cut for it in a single place, and that every
mechanical loader on the market can load coal
very much faster than it can be hauled away.
"It is the fact that they must in the nature of
things"—or rather in the nature of unmodified
room-and-pillar mining—"be peripatetic, wandering about from place to place, or else they
must be idle, and sometimes"—under the ordinary methods of mine transportation—"that
they must be both peripatetic and idle, which
makes their introduction slow and halting."†
But vigorous attacks are now being made on
both of these problems. To do away with the
waiting for cars, present-day methods of transportation are being further standardized; double

*Reported at 1924 Mining Congress.
†Hall *op. cit.*

track or y-shaped track-ends are being placed in
the rooms so that the machine can fill one car
while the other is being shifted; and in some
cases more elaborate expedients are being de-
vised to make it possible to bring whole trips
of cars past the discharging boom of the conveyor. At least two companies, moreover, have
carried to completion the process foreshadowed
in the account of the Norton mine and have
combined machine loading with conveyor haulage in a complete mechanical system.* But
most of these changes are difficult to arrange,
and the more drastic and the more effective of
them are almost impossible, in the narrow rooms
of ordinary mining; and this fact, together with
the equally urgent demand for working places
large enough so that the machine need not
"wander" from one to another, is leading a number of operating men to experiment with changes
in the lay-out of the mines. One result of this
is quite naturally a re-examination of the possibility of winning the coal by the longwall
method—in which all the coal is taken out in
one continuous process on long straight faces—
a method which is common in other countries
but rarely practised in America except in low
coal. A number of experimenters have been
attempting to adapt this system to American

*Coal Age, July 24, 1924.

conditions without the almost prohibitive expense of "backfilling" the mined-out areas to support the roof; and at least one company has been confident enough of success to develop a double conveyor machine built expressly for crawling down the longwall faces. In the meanwhile, a number of other ingenious schemes have been worked out with the aim of obtaining the advantages of long faces and easy haulage by means of less drastic modifications of room-and-pillar mining,—by *sidewall* mining, for example, in which the machine takes long "slabbing cuts" down the sides of an ordinary room and by various adaptations of the system of "saw tooth pillar points" already discussed;* and it is perhaps in this direction that the most active development is to be expected in the near future.

Even without the more drastic of these changes, the number of machines in use will no doubt continue to increase. There are still many obstacles to be overcome, to be sure, and difficulties in cleaning the coal from certain seams; but even to-day gatherings of operators do not laugh any too confidently when a manufacturer tells them that they'd better get out of the business if they can't use loading machines

*Walter M. Dake, "How to Get Big Output From Loading Machines," *Coal Age,* May 8, 1924.

in their mines, for there are enough that can to supply all the country's needs. And if by any method, the machines can be assured continuous operation throughout the working day, there can be no doubt at all of their rapid spread and of the rapid disappearance of the old indiscipline. "Give the loading machine, the face conveyor, and the electric drill steady work," writes an enthusiastic advocate of the change, "and flesh and blood shall not prevail against them."*

But what *is* happening to "flesh and blood," and to the life of the miners, as this revolution goes on? Many different things, of course, in a movement so many-sided; but at least something of the main directions of the change may be made out from the description. In the first place, it means of course, the passing of the hand shovel, and with it, as the hand loader goes the way of the old-time miner, the elimination of the group of workmen whose customs and traditions have done most to determine the character of the life within the mines. And it means also the elimination of the institution of the place and with it some of the most striking elements of the old indiscipline, for the curious bundle of rights that go with the miner's yearlong possession of a particular room will surely never be put together again under the machine. The most

*Hall, *op. cit.*

characteristic of the old jobs and the most characteristic set of working arrangements are thus done away with; what of the new jobs and what of the types of discipline that are taking their places?   The answer most commonly given is that the change will put a premium on intelligence and replace dull and heavy manual labor with jobs that consist mainly of the skilled manipulation of control levers.   Under the new system, declares the Coal Commission confidently, "the miners left will be of a higher type," and the work to be done will call for "greater initiative" than before.*   "If the loading machine does away with the old back-breaking shovel," says an Indiana operator, "naturally the man who graduates to it from the hand shovel will like the change."†   But this answer needs to be checked carefully against the list of the new occupations.   Hand loading, to be sure, is irksome and arduous enough, and it is quite true also that the actual operation of the mechanical loader does frequently call for a considerable degree of skill.   "The same class of men that make good (cutting) machine runners," says a Jeffrey advertisement, "make good men for the mining and loading machine;" and occasionally the job is considered even more

*Thompson Report, p. 6.
†Ingle, op. cit.

highly skilled than those of the cutters and motormen. But this is by no means always the case. A certain superintendent who was in charge of the work of several Holmested machines scornfully denied that he was in the habit of taking cutting machine runners to train for the mechanical loaders. "Why should I?" he demanded. "Cutters make too much money." One company even claims that "any mine laborer can learn to run the Shuveloder in two hours;" and the tendency both in designing and in advertising the machines is clearly in the direction of emphasizing simplicity of operation. Even with the slightly more complicated makes, the sales engineer who remarks that the machine is going to "elevate the miner's work" is quite likely to claim in his next sentence that he could take you out and teach you to run it in twenty minutes. The comparison, moreover, must not rest merely with the operating job. For every man who "graduates" to machine-running, there is also one man who "graduates" to the almost entirely unskilled job of car shifting; and on the average there is still another who is transferred to a new job at the face which does require something of the craft knowledge of the ordinary miner—enough at least to keep him from being buried by the coal that he picks down—but which is much less varied and surely calls for much less "initiative" than the work in a separate room. On the other hand, the separate drilling and shooting add new and quite highly skilled jobs to the list of mine operations that are performed by roving crews—at least until the cutters-and-loaders do away with them again; and the handful of men that are needed to keep the new machines in repair must of course be skilled mechanics.

On the point of discipline and supervision, however, there is no such doubt of the direction of the change. Machine loading "means that the foreman, instead of making an occasional visit to the miner, is in constant touch with all of the men," declares the Thompson Report; and its authors give this possibility of keeping both the machine gangs and the other workers "under such definite control" as a chief advantage of the introduction of the machines.* In some cases, to be sure, the increase in supervision is relatively slight. With the smaller machines operating under an unmodified room-and-pillow system, the skilled operator and his one or two helpers may preserve a good deal of the older independence, particularly if the piece work basis of payment is retained; but always there is the fact of concentration—of increased tonnage from a given area—to make supervision

*Pp. 12, 13.

easier, and the fact of the investment in the
machine—with the company the very definite
loser by its idleness—to make supervision more
obviously profitable. The machines do not lay
off; "A Shuveloder," so an advertisement runs,
"doesn't lubricate on 'moonshine whiskey' . . .
it doesn't kick about bad air, heat or cold . . .
it loads as much at 4 P. M. as 8 A. M.;" and the
men who attend the machines are expected to
imitate their regularity. Even scraper loaders
are in some cases operated under a rigid gang
control; with the smaller machines of the double
conveyor type, manufacturers are sometimes ad-
vocating and operators frequently adopting a
system under which there is a boss over the cutting, loading and auxiliary work required for a
single machine unit—one boss over some ten
men, that is, instead of over a hundred; and
with the Coloder there is a boss in constant
charge of the machine gang itself. Each adoption of the cutting-and-loading machines, moreover, and each change to conveyor haulage,
makes a further drastic cut in the number of
scattered and slightly supervised workers; and
every one of the current changes in mine lay-out
is bringing with it a nearer approach to the factory grouping of labor.

In still another way mechanical loading has
led towards the new discipline. In most cases

it has meant day work, and day work means an increase in bossing. The machines, to be sure, were neither designed nor bought for this purpose. "If the manufacturers have been telling you that mechanical loading is likely to do away with piece work," declared a prominent coal man in 1923, "they'd better not tell the operators or they won't sell any machines." Yet their actual use has been almost entirely on a day work basis; and, although many operators think of this as merely a temporary expedient until the capacity of the machines is determined, *Coal Age* is now declaring that "machine loading is not piece work in the nature of it"* and it is quite possible that the change which began as sheer by-product may become the settled policy of the industry.

Whatever its effect on the method of payment, there can be no doubt that the road of mechanical loading is a road toward the new discipline. Just now, moreover, it is the most-traveled road, and in many ways it is the most far-reaching. To be sure the many-sided process that is bringing the machines into the mines and modifying the mines to receive them has not

---

*July 24, 1924. The editorial from which this is taken welcomes the day wage scale just signed in Illinois as a great victory for the operators and predicts that "it is the die in which all future wage agreements for machine operation probably will be cast." But see also the debate in the issue of Aug. 21, 1924.

been guided by any such clear aim of increasing supervision and decreasing individual initiative as is the policy of the United States Coal and Coke Company; nor does it give rise to any group of jobs that are both so repetitive and so completely bossed as those of the men who shovel onto the face conveyors; but unlike the first it bases its standardization on the capacity not of men but of machines, and it goes even farther than the second in the actual mechanization of the work of the industry.   Each separate innovation cuts away one of the supporting elements of the old indiscipline, and the resultant of them all is a powerful tendency toward the greater regimentation of work.

This movement, moreover, is beginning to find something of a concerted policy in a number of programs for the complete mechanization of the miners.   And when that is accomplished, runs the exultant phrase, "the man won't be a miner any more, coming to work when he pleases; he'll be a mechanic"—or an unskilled laborer—"and he'll do what he's told."   The recent Giant Power Report declares that "mass production in the fullest sense of the term . . . 'Ford methods' in short . . . must be practiced all the way from the face of the coal seam;"* and there is a growing body of opinion

*Report of the Giant Power Survey Board to the General Assembly (Pennsylvania).   Morris L. Cooke, Director.   P. 42.

within the mining industry that agrees with its authors. Even the most "progressive" mine of to-day is very far indeed from the extreme of the "Ford methods," but one prominent superintendent declares that just such a degree of regimentation can be attained as soon as a sufficiently automatic cutter-and-loader can safely be driven down a longwall face, and another argues for a standardization of "every operation down to the minutest detail so that no responsibility of any kind will fall on the worker."* The present-day contrasts are by no means so sharp and dramatic, yet the direction of the change is clear and unmistakable. The new jobs, whether skilled or unskilled, are most of them performed under a supervision and a control more like that of the factories than like the older indiscipline of the mines. "The last link in the machine-run mine is being forged," reports *Coal Age,* and adds as an obvious corollary that "factory methods, henceforth, will rule in the operations of the industry." A typical headline in the same journal declares that one of these changes "Raises Mine to Factory Level;"† and a perhaps equally typical union committeeman describes a certain highly mechanized mine as "nothing but a God damn factory." "It's a coal

*Stroup, *op. cit.*
†April 24 and May 29, 1924.

factory, that's what it is," he said; and the description without the expletives would no doubt have been cheerfully accepted by the company; for both friends and foes of the movement would agree that it means the introduction, and the fairly rapid introduction, of factory methods and the factory psychology within the mines.

PART III

Conclusion

CONCLUSION

CHAPTER VI

THE PROBLEM OF THE WORKING LIFE

THE unusual independence of the miners has been rooted in the peculiar technique of the industry and in their own firmly-held traditions. The onward march of the machine process is doing away with this older technique and pointing to a sharply different sort of life and discipline. What, then, is to be the future of the miner's freedom?—It is this problem that the present study is intended to raise.

So far, to be sure, the clash of the two types of life has been by no means a spectacular one. There has been no sudden and widespread introduction of stop watches and time clocks to mark the change, and no dramatic burning of the picks and augers and shovels of the old technique. The revolution is still in its early stages, and its more extreme forms have so far forced their way only into a fraction of the industry. The actual introduction of mechanical loading, in fact, has hardly outrun the rival and parallel innovation of *stripping*,—which turns the mines not into factories but into quarries and makes mining a different and an outdoor industry by the bold stroke of lifting the entire cover off the seams of coal; and, although the change described in this book is no doubt of much greater significance for the future of the industry, the three hundred or more loading machines that are its most tangible expression are scarcely producing more coal than the one or two per cent of the country's output that is supplied by the steam shovels of these open strip pits.

This march of the new technology into the mines, moreover, has been confined as yet very largely to the particular regions in which the traditions of the miner's freedom have been the weakest.   The revolution has so far been largely a non-union one.   The company with the tri-colored absence slips and the mines that work under the "estimate" system are well beyond the reach of union organizers; the splitting up of the loader's job has been mainly a non-union movement; and, although the West Virginia Coal and Coke Company started its V system under union auspices, the men who shovel into the face conveyors at Norton to-day are for the most part farmers from the neighboring hills who have had no experience either with mines or unions.   With the loading machines, to be sure, the case is somewhat different.   The two leading commercial manufacturers say that it is easier to sell machines to union operators than to non-union ones, for the latter can cut costs by cutting wages instead, and the machines in actual use seem to be fairly equally divided between union and non-union fields. But even here, the records of the largest outputs and the experiments with the types of machines that carry the new discipline the farthest are almost entirely non-union. Nor is this the whole story. The men who work under the new technique are in many cases not merely non-unionists but also men who have had little or no exposure to the traditions of the gob-pile and of the industry. "A hopeful characteristic of the unskilled miner," it is said, "is his ready surrender to discipline;"* and boss after boss in the more modernized regions echoes this reason for preferring "greenhorn" labor. The separateness of the working force of the United States Coal and Coke Company has already been referred to; and its preference, and that of its allied companies, for foreigners fresh from Ellis Island is well known. Certain of the Southern regions in which the new discipline has gone the farthest, moreover, have other great supplies of inexperienced workers to draw upon,—mountain whites from the isolated creek valleys that have been suddenly invaded by the "Black Avalanche"† of coal mining, men that are "the tim-

---

*"Problems of Operating Men," *Coal Age,* July 21, 1921.
†Winthrop D. Lane, *Survey,* Vol. XVII, p. 1002.

ber that lived (there) when a mine was an unheard-of thing," and negroes escaping from the
low wages of the lowland farms—"cornfield
niggers," as the industry knows them. "What
did you work for down there?" one of the latter
was asked. "For nothin', boss!" he replied, and
his grin and his manner suggested the possibility
of a quite cheerful "surrender to discipline." It
is with workers like these that many, though
not all, of the experiments with the new technique have been made; and the craftsmen in the
solid-shooting mines of Southern Illinois, for
example, are scarcely aware that the change is
going on. So far, in fact, the two types of life
have hardly met in open conflict.

This relative isolation, however, can hardly
continue for long. No one expects the new
technique to remain confined to the present scattered experiments. The rapidity of its spread,
to be sure, may well be a matter of doubt and
debate. Any success in the current attempts to
regularize the industry and lift it from its present two-hundred-day-a-year status would of
course greatly hasten the process; but any
further overdevelopment of the industry, such
as the machines themselves may well stimulate,
would delay it; and those who see the future of
the process as one of successive rapid advances
and sharp set-backs are in all probability better

students of the economics of coal than those who predict a steady and uninterrupted increase of the new technique.* But surely no observer of the savings already effected by the machine methods, and no student of the general trend of modern industrial development, can doubt either their further intensification or their ultimate extension to the rest of the industry.

New methods will surely brush aside the old technique; will they brush aside also the qualities of independence and initiative which that technique has fostered? The answer cannot be made in terms of mechanical processes alone. The freedom of the miners has been a matter of their own activities also, and its future will in part depend on how their union meets the new technique and on how successfully it weathers the difficulties of the transition. Already its attitude is the subject of vigorous controversy. Now as always, declare the spokesmen of the operators, the organization stubbornly opposes technical progress. Now as always, declares the union president, "the United Mine Workers favor the introduction of labor-saving machinery." And now as always, also, the truth lies somewhere between the two categorical statements.

*Walton H. Hamilton and Helen R. Wright, *The Case of Bituminous Coal*. Institute of Economics (to be issued shortly).

"It is a well-known fact," testified a leading operator, "that the miners as a general rule oppose the introduction of labor-saving devices or machinery in the mines." "Whether it comes . . . from force of habit or a selfish motive," he added, "is immaterial to the operator."* For this familiar charge there is a considerable factual basis. New machinery is of course introduced by and for the employers, whatever the final gains of the consumers may be, and certainly neither by nor for nor even—except as the union forces it—in consultation with the workers; and it is not surprising that they are less interested in mechanical progress than the operators. "I never was a believer much in machinery," a local leader once remarked, "yet I never was one that took a stand against (it). . . . But I always believed that in introducing any machinery . . . the workers ought to be considered as well as the employers."† Certainly it is true that the miners, like other workers with well-established traditions, have on the whole been reluctant, both from "force of habit" and "ignorant prejudice" and pride of craft, to submit to changes in their methods of work; and it is true that workers who have seen that the first effect of the introduction of machinery, what-

*Bituminous Coal Commission, *op. cit.*, Vol. XII, p. 651.
†District 2, *Joint Conference Proceedings*, 1914, pp. 40-1.

ever the ultimate gains and losses, would be that of throwing men out of work have often opposed it from the "selfish motive" of wanting to keep their jobs. As a result many groups of miners, with or without the sanction of their local unions, have from time to time flatly opposed technical changes. There was "a contention amongst the men," as one union official put it, "at most of the mines when the (cutting) machines were first put in;"* there is a contention to-day at many a mine where loading machines are being introduced; and it is not difficult to pick up stories of quiet but successful sabotage against the cutting machine in the older days or even of an occasional proposal in more recent times to dynamite a certain new piece of machinery. The official policy of the union has been a quite different one, however, and the international organization, from the time of John Mitchell to the present, has strongly opposed such outbreaks. "We have always favored such machinery," declares President Lewis in a vigorous restatement of the traditional official viewpoint. "The one thing the organization insists on is that the mine workers be given their fair proportion of the benefits."† Much the same stipulation, moreover, was made

*District 2, *Joint Conference Proceedings*, 1912, p. 396.
†Quoted in "News of the Industry," *Coal Age*, Jan. 31, 1924.

in Section 3 of the award of the Bituminous Coal Commission in 1920 and has been written into the joint agreements that have followed it. "The right of the operator to introduce any such new device or machinery shall not be questioned." During the experimental period, he shall pay wages "at least equal to the established scale for similar work." When that is past, a joint test of the performance of the machine shall be made to "disclose the labor-saving;" and then a scale is to be drawn up which will give to the workers "the equivalent of the contract rates for the class of work displaced, plus a fair proportion of the labor-saving effected." This policy and this provision have at least the effect of taking the matter out of the realm of flat rejection and into the realm of bargaining over the terms of acceptance. They do not of course take it out of the realm of controversy. It is a common remark at operators' gatherings that "John L. Lewis's idea of a fair share is a hundred per cent;" and there are of course many men on the other side who take the point of view—so long conventional in economic theory—that the workers' fair share, as workers, in the gains of technology should be nil. The result has been a great "higgling of the market" and, as in other markets, there have sometimes been irritating delays and occasionally there have been complete failures to agree. It is frequently claimed, moreover, that the rates adopted have themselves discouraged the introduction of the new machinery. It is charged that "the union has compelled the operators who provided mechanical devices for the mining of coal to accept such small differentials as to leave in many cases little or no profit in the use of such equipment."* It is at least true that the effect of the union's policy has been to leave *less* profit than there might otherwise have been; but nevertheless it is still an open question how far all this has actually retarded technical changes. It is certainly not the striking factor that has distinguished coal mining from other industries, and in the case of the most disputed of past changes, the introduction of the cutting machine, there is little or nothing in the figures of machine-introduction in various fields to suggest that this retardation has been marked enough to give the non-union fields any advantage in the use of machinery. The careful analysis made by Mr. Lubin for the Institute of Economics† in fact concludes that "the machine differential has not up to the present time seriously interfered with the rapid adoption of mining machinery." "In the near

*R. Dawson Hall, "Coal-Mining Technique Progressed Greatly in 1921," *Coal Age*, Jan. 18, 1922.
†I. Lubin, *Miners' Wages and the Cost of Coal*, Ch. XIII, Sec. I, esp. pp. 260-3.

future," however, as he points out, "the system of machine differentials" may well "become an issue of paramount importance." It will certainly be one of "the important factors in determining the rate of installation" of new machines, and it will be an even more decisive factor in determining the significant relation between the rates of their introduction in union and non-union fields. "Which section of the country will complete the mechanization of the mines first?" asks *Coal Age*. "Will the fear of losing markets . . . so spur the union operators that they will put in conveyors and loading machines? Or will the non-union operators out of their plethora of profits, present and accumulated, and their shortage of mine labor lead the way?"* In the past changes the union operators have apparently kept at least abreast of their competitors in the matter of machinery, and even to-day there is no great difference in this respect between the ordinary union and the ordinary non-union mine. But it is a significant fact that the absolutism of non-union operation has given to a minority of operators the opportunity to make the greatest advances on the road of the new discipline, and "the remarkable success which is attending the mechanization of

*May 15, 1924.

non-union mines" is already being pointed out as a "serious portent for the union fields."*

The question of what the machine will do to the union, in fact, is quite as important for the present study as the question of what the union will do to the machine; and there are even a number of observers who believe that "the changes now taking place in coal mining systems will eventually result in the elimination of the United Mine Workers." One Western mine superintendent, in fact, Mr. Thomas A. Stroup, expressly argues for the complete mechanization of the mines as a certain cure for unionism; and he bases his confidence in its effectiveness on the sound observation that for the most part unions flourish where the type of work is still largely individualistic—as on the railroads and in the hand trades—rather than where repetitive jobs and standardized labor methods prevail. Certainly it seems to be true that the men who work under "mass production systems" are "less susceptible" than other workers to what the writer calls the "corrupting influence of unionism"—although their lack of skill is a more obvious explanation than the lack of grievances that Mr. Stroup assumes; and it is possible to cite striking cases in which the

*Paul Wooton, "High Cost Operators Panicky at Likelihood of Long-Time Wage Agreement," *Coal Age,* Feb. 21, 1924.

change from individual to routine work has been followed—as in the metal mines and the steel mills—by the weakening or break-up of once powerful unions.*   The argument is at least an arresting one; and it is not to be brushed aside too lightly even in the case of an industrial union that is as powerful and aggressive as the United Mine Workers and that is so firmly rooted in the whole life of the mining villages throughout the greater part of the industry. It is hard to imagine any set of changes that would break down the almost religious loyalty of the workers toward their union in many of the older fields, and almost impossible to imagine its elimination so long as the industry continues to be manned by much the same personnel and carried on in much the same regions as the present time; but it is quite possible that there may be a shifting in these very respects and that the industry of the future may be based much more largely than at present on the productivity of the greenhorn workers and the non-union mines of certain of the newer fields. Already the new technique beyond the union border is a factor in taking work away from union miners; and if the organization should retard its advance too long, or bargain too stubbornly with the new machines,

---

*Stroup, *op. cit.* This article is discussed in a *New Republic* editorial (by the present author) in the issue of Jan. 30, 1924.

it might find the industry itself moving bodily away from the union and its loyal membership.

This threat from the non-union fields would stand as a definite check upon any attempt to formulate a policy for preserving the miner's freedom; and so far, moreover, there is little in the tenor of the union publications or in the day-to-day talk of union offices to indicate that the major problems of the new technique are under very active discussion. One high international official not long ago expressed confidence that the superior craftsmanship of the organized miners would be an important offset to any advantage the non-union fields might gain by the new methods—though their effect is apparently to make just that craftsmanship useless; and the international organization has left the immediate issues of the change almost entirely in the hands of the district unions with little more than a reiteration of the traditional policy that was worked out when the problem in hand was the quite different one of the coming of the cutting machine. At least one conference of district representatives, to be sure, has discussed the question of a scale for loading machines; and the powerful organization in Illinois is showing its awareness of technical developments by calling in expert assistance to study the possibilities of Giant Power in the hope that a change that promises less work for the industry as a whole may by foresight be made to promise more work for *Illinois* mines. But for the most part the problems of the new technique have not yet forced themselves upon the full attention of the miners and their representatives; and when they do so, it is possible that all questions of the quality of the working life may still be put aside in the urgent concern over the issues of wages and jobs. For the actual battles will for the most part be fought over the rate of wages for the men who work with the machines; and the most compelling consideration, both to the rank and file who will see the machine primarily as a device for taking away their jobs and to the leaders who recognize that the industry is already overmanned as well as over-mined, may well be its effect on the problem of employment. The union's main concern, indeed, is much more likely to be with the problem of keeping any jobs at all for its men than with that of determining too narrowly their exact nature. Yet in spite of all this, there are certain indications, both in the attitude of the rank and file and in the activities of the organization, that the traditional freedom of the miners will not be given up without something of a struggle and even that there may be a definite attempt to win new types of freedom as the older type becomes obsolete.

One sign of this is the union's insistence, so far unsuccessful, that the loading machine men be paid on a tonnage basis. There are of course many other reasons for this attitude,—the fact that coal always has been loaded by the ton, for example, and the belief that tonnage rates would give the possibility of higher earnings; but it is clear from the talk of local leaders in some of the disputed cases that one genuine motive has been the fear that the independence traditionally associated with the contract work would be lost in the change. Payment by the day, they feel, means a greater likelihood of the boss looking down the shirt collar. The same attitude, moreover, came out even more clearly in the union's suggestion for the method of payment with the face conveyor. Ordinary piece work, of course, is virtually impossible since the conveyor does not keep separate the individual loadings, and the company naturally put the work on a day basis. The union, however, proposed a piece work payment based on the tonnage of the whole group of shovelers and divided equally among them. To the company officials the notion seemed preposterous. "Would you or any man go in there and work under that system?" they demanded; but certain of the spokesmen for the union argued that it would be perfectly practical. But how would you make a

lazy man work if his earnings depended mainly
on the work of the others and not on his own?
"Well, this way. Suppose a fellow stopped to
light a cigarette. The fellow next to him would
say, 'We're not smoking to-day, buddy, we're
loading.'" What if he still wouldn't work?
"The men next to him would speak to him two
or three times as United Mine Workers; and if
he kept on loafing they would have the mine
committee tell the foreman not to fire him but
to put him on some other kind of work." "That
way," they added, "you wouldn't have to have
that boss in there all the time." The proposal
was an isolated one, to be sure, and was not
pushed very far even in the particular case, but
it indicates a significant attitude on the part of
its advocates toward the possibility of freedom
under the changed technique. Even if the new
device is to take away the independence that
went with the old-time isolation, so their argument seems to run, it need not mean that the
workers must be watched all day by a straw
boss.

Group action of the sort suggested, moreover,
is not entirely without precedent or background
in the industry. The advocates of the plan
were able to point to other cases in which the
union had assumed responsibility for certain
matters of discipline, and the ordinary activities

of hundreds of mine committees on the wide range of questions discussed in Chapter II often give to the workers who turn to the committee-men with their demands an important sense of group power and freedom. Something at least of the independence described in this study will remain characteristic of the industry—for better or worse—as long as the union and the mine committee survive; and there is within the union at least the beginning of movement toward the extension of this type of workers' control from its present negative form to a much more responsible part in the running of the mines. This point was a conspicuous one, for example, in the pamphlets published under the union's authority by the Nationalization Research Committee. "Democratic management," they declared, "must be the blood and bones of a plan for nationalization;" and they made it clear that they meant by this something closer to the life of the ordinary worker than the mere election of union representatives to serve on some national board of management. The present sort of control that is exercised by the mine committees has given to the miner "real gains in personal freedom and economic status," they said, but it is "not enough for a worker in a democratically managed industry . . . The working miner must have a real part in the government of coal."* And although the program of workers' control as stated in these terms has appealed neither to the strategic sense of the union officialdom nor very strongly as yet to the imagination of the rank and file, and although the whole idea of nationalization as an immediate policy was quietly laid aside at the 1924 convention, there is a continual pressure among active unionists toward increasing the power and influence of the mine committees. The clause affirming the exclusive authority of the operator in matters of management is a real subject of contention, and the union official who met its reading in an arbitration case with the stout declaration that "there ought to be a limit to any right he has got"† was expressing a widespread feeling among the rank and file. The union already exerts a considerable group control; there is within it a demand for more control, and it may be that this demand will be greatly increased as the new methods break down the individualism of the scattered miners and throw them together into closer groups. For if the unrest of a few ex-miners under the new

*Nationalization Research Committee, (John Brophy, Chairman, C. J. Golden, William Mitch), *How to Run Coal,* pp. 10-11. See also *Compulsory Information in Coal* issued by the same committee; John Brophy, *Facts!*; and the two pamphlets, *Why the Miners' Program* and *The Government of Coal,* issued by District 2.
†James Mark, Vice-President of District 2. In *Arbitration Proceedings,* April 15, 1919, p. 160.

discipline out in the factories has been conspicuous enough to be noticed by a number of observers, the coming of factory methods into the very strongholds of the miner's freedom may well be expected to lead to even greater restlessness and discontent. And as the onward sweep of the machine process carries away the old indiscipline of the mines, that discontent may take the form of a movement toward what the operators would doubtless call the *new indiscipline* of increasing workers' control,—a demand, that is, for a miners' freedom to take the place of the miner's freedom they are losing in the change.

In spite of the danger confronting it, then, and in spite of its pressing concern with other problems, it is still possible that the union may exert a powerful and distinctive influence over the quality of the working life under the new technique; but so far the reports of its discussions give little indication of a conscious attempt to control the change in terms of such considerations. Nor is there much evidence of concern over the life on the job among those who are planning and initiating the new technique. There is an occasional flurry of comment, to be sure, over the suggestion that the new methods mean the elimination of individual responsibility. One enthusiast declares that this is precisely as it should be; the effect will be to make the workers happier and more contented. "The methods of Ford are not only economically sound, they are socially and psychologically sound and must be extended to all industry." Other leaders in the technical revolution, however, deny with equal vigor that the change means any narrowing of the workers' task. One company announces that it has introduced mechanical loaders "not as has been suggested to make 'mindless machines of its employees' but instead to translate the job of producing foot-pounds with a shovel into that of operating a machine where a man can use his brains more and his back less."* And a prominent general manager, who is as impatient as any with the inefficiencies of present-day mining, declares stoutly that he would rather go out of the business than attain efficiency by a system under which the bosses did all the thinking for the men. Yet such discussion is rare, and there is little evidence that these considerations are playing any significant role in the decisions that are turning the mines into factories. For the most part the revolution of conveyor and loading machine, like that of steam engine and spinning jenny, goes on quite without trace of the intention to make men's jobs either better

*"McAuliffe Increases Safety in U. P. Mines; Installs Loaders," *Coal Age,* Dec. 20, 1923. See above, pp. 148-150.

or worse, or to make the life upon them either freer or more rigidly controlled.

The ruling consideration, both in the minds of the operators and in the minds of those who are studying the change from the viewpoint of the consumers, is of course the expectation of lower costs and the chance to turn them to account in higher profits or in lower prices. "On to Cincinnati for Lower Costs Per Ton," was the appropriate slogan for the meeting at which the innovations were most thoroughly discussed. The technicians and those who are actually installing the new devices add to this motive, no doubt, a delight in the skillful planning and the nice adjustments for which they call. Other keen observers are watching the change with an eye to its effect on the problem of overdevelopment—some of them fearing the effect of so sharp an increase in the capacity of an industry that needs nothing quite as much as a concerted decrease in capacity, and others hoping that the greater equipment required by the new technique may make it harder for new operators to pour into the industry at every flurry of demand. Still others are asking the effect of the changes on the conservation of the coal resources or on the possibility of the better utilization of coal under the Giant Power and other projects. The miners themselves, as has been shown, are likely to see the new technique mainly as a threat to their jobs and to their wage rates. Other observers, however, point to it as a promising source of increased pay and better living conditions. And surely its economies are marked enough to make such an increase at least possible, and quite probable if the change does not at the same time weaken or destroy the workers' chief agency for asserting their claims. On the score of safety, moreover, the advocates of the new methods hold out even more glowing promises. Falls of rock are a matter of time, it is pointed out; the great speed of the new machinery will get the coal out before the roof has a chance to come down; and the new supervision will add a great safeguard against human carelessness.* And although this assurance is somewhat lessened by the simultaneous claims that these things will also make possible the use of less timber and the winning of coal by methods that have been previously considered too dangerous, here again is at least the possibilty of a great gain for the men in the mines.

Profits and prices, and to a less extent employment and wages and safety,—it is chiefly in these terms and the others suggested above that the revolution in the mines is discussed; and certainly all these things should be con-

*Mason, *op. cit.* p. 114. *Thompson Report,* p. 13.

sidered as men weigh the gains and losses of the change and attempt to control its course. But for all these claims there are already spokesmen, and for most of them—though for some much less perfectly than for others—channels are already provided by which their consideration does actually affect the choices of the industry. The purpose of the present study, on the other hand, is to place alongside these other problems an almost entirely neglected one,—that of the effect of the change on the quality of the working life. For the old technique of the mines, with all its other advantages and disadvantages, has given to the workers an independence quite unique in industrial life; and in just this respect the revolution brings with it some of the sharpest and most radical changes. The question, therefore, of what sort of working arrangements are to take the place of the old,—of how far the new discipline is to destroy the individualistic freedom of the miners and of how far that older freedom may be replaced by the quite different collective freedom of workers' control—is of vital importance for the life of the mining population. It is for this problem that the present book asks consideration, and a consideration based neither on the glorious fanaticism of a William Morris who declares that it were better that the heavens should fall than that any work should be a torment to the worker* nor on the uncritical modern fatalism of those who take it for granted that work must always and inevitably be dull. For ways of working are still a significant part of men's lives; and the revolution that is beginning to change the stubborn individualists of the gob-piles into "routine workers whose tasks are minutely set and closely supervised"† is by no means the least striking of the technical changes that have been turning men from beasts of burden into crane operatives and from craftsmen into feeders of machines.

*William Morris, *Signs of Change,* p. 172.
†Stroup, *op. cit.*

INDEX

WORK

Its Rewards
and
Discontents

An Arno Press Collection

Anderson, V.V. **Psychiatry in Industry.** 1929

Archibald, Katherine. **Wartime Shipyard.** 1947

Argyris, Chris. **Organization of a Bank.** 1954

Baetjer, Anna M. **Women in Industry.** 1946

Baker, Elizabeth Faulkner. **Displacement of Men by Machines.** 1933

Barnes, Charles B. **The Longshoremen.** 1915

Carr, Lowell Juilliard and James Edson Stermer. **Willow Run.** 1952

Chenery, Wiliam L. **Industry and Human Welfare.** 1922

Clark, Harold F. **Economic Theory and Correct Occupational Distribution.** 1931

Collis, Edgar L. and Major Greenwood. **The Health of the Industrial Worker.** 1921

Dreyfuss, Carl. **Occupation and Ideology of the Salaried Employee.** Two vols. in one, 1938

Dubreuil, H[yacinth]. **Robots or Men?** 1930

Ellsworth, John S., Jr. **Factory Folkways.** 1952

Floyd, W.F. and A.T. Welford, eds. **Symposium on Fatigue *and* Symposium on Human Factors in Equipment Design.** Two vols. in one, 1953/1954

Friedmann, Eugene A. and Robert J. Havighurst, et al. **The Meaning of Work and Retirement.** 1954

Friedmann, Georges. **Industrial Society.** Edited by Harold L. Sheppard. 1955

Goldmark, Josephine and Mary D. Hopkins. **Comparison of an Eight-Hour Plant and a Ten-Hour Plant.** 1920

Goodrich, Carter. **The Miner's Freedom.** 1925

Great Britain Industrial Health Research Board, Medical Research Council. **British Work Studies.** 1977

Haggard, Howard W. and Leon A. Greenberg. **Diet and Physical Efficiency.** 1935

Hannay, Agnes. **A Chronicle of Industry on the Mill River.** 1936

Hartmann, George W. and Theodore Newcomb, eds. **Industrial Conflict.** 1939

Heron, Alexander R. **Why Men Work.** 1948

Hersey, Rexford B. **Workers' Emotions in Shop and Home.** 1932

Hoppock, Robert. **Job Satisfaction.** 1935

Hughes, Gwendolyn Salisbury. **Mothers in Industry.** 1925

Husband, Joseph. **A Year in a Coal-Mine.** 1911

Kornhauser, Arthur, Robert Dubin and Arthur M. Ross, eds. **Industrial Conflict.** 1954

Levasseur, E[mile]. **The American Workman.** 1900

Lincoln, Jonathan Thayer. **The City of the Dinner-Pail.** 1909

Man, Henri de. **Joy in Work.** [1929]

Marot, Helen. **Creative Impulse in Industry.** 1918

Mayo, Elton. **The Human Problems of an Industrial Civilization.** 1933

Mayo, Elton. **The Social Problems of an Industrial Civilization.** 1945

Moore, Wilbert E. **Industrial Relations and the Social Order.** 1951

Morse, Nancy C. **Satisfactions in the White-Collar Job.** 1953

Myers, Charles S. **Industrial Psychology.** [1925]

Noland, E. William and E. Wight Bakke. **Workers Wanted.** 1949

Oakley, C[harles] A[llen]. **Men at Work.** 1946

Odencrantz, Louise C. **Italian Women in Industry.** 1919

Purcell, Theodore V. **Blue Collar Man.** 1960

Reiss, Albert J., Jr., et al. **Occupations and Social Status.** 1961

Reynolds, Lloyd G. and Joseph Shister. **Job Horizons.** 1949

Rhyne, Jennings J. **Some Southern Cotton Mill Workers and Their Villages.** 1930

Richardson, F. L. W. and Charles R. Walker. **Human Relations in an Expanding Company.** 1948

Roe, Anne. **The Psychology of Occupations.** 1956

Sayles, Leonard R. **Behavior of Industrial Work Groups.** 1958

Seashore, Stanley E. **Group Cohesiveness in the Industrial Work Group.** 1954

Shepard, William P. **The Physician in Industry.** 1961

Smith, Elliott Dunlap. **Technology & Labor.** 1939

Soule, George. **What Automation Does to Human Beings.** 1956

Spofford, Harriet Prescott. **The Servant Girl Question.** 1881

Staley, Eugene, ed. **Creating an Industrial Civilization.** 1952

Stein, Leon, ed. **Work or Labor.** 1977

Suffer the Little Children. 1977

Tead, Ordway. **Human Nature and Management.** 1929

Tilgher, Adriano. **Work:** What It Has Meant to Men Through the Ages. 1930

Todd, Arthur James. **Industry and Society.** 1933

Tugwell, Rexford G. **The Industrial Discipline and the Governmental Arts.** 1933

U.S. House of Representatives, Committee on Labor. **The Stop Watch and Bonus System in Government Work.** 1914

Vernon, H[orace] M[iddleton]. **Industrial Fatigue and Efficiency.** 1921

Walker, Charles Rumford. **Steel:** The Diary of a Furnace Worker. 1922

Whitehead, T.N. **The Industrial Worker.** Two vols. in one. 1938

Whitehead, T.N. **Leadership in a Free Society.** 1936

Whyte, William Foote. **Human Relations in the Restaurant Industry.** 1948